THE FUTURE
OF PROGRESS

Reflections on Environment and Development

THE FUTURE
OF PROGRESS

Reflections on Environment and Development

EDWARD GOLDSMITH, MARTIN KHOR
HELENA NORBERG-HODGE, VANDANA SHIVA
& OTHERS

Published by Green Books
in association with
The International Society for Ecology and Culture

A Resurgence Book
published in 1995 by
Green Books Ltd
Foxhole, Dartington
Totnes, Devon TQ9 6EB, UK

in association with

The International Society for Ecology and Culture
21 Victoria Square, Clifton, Bristol BS8 4ES, UK
and PO Box 9475, Berkeley CA 94709, USA

This is a revised edition of the book of the same title
published by ISEC in 1992

Edited by
Helena Norberg-Hodge, Peter Goering and Steven Gorelick
Typesetting by Chris Fayers, Soldon, Devon
Cover design by The Big Picture, Bristol
Printed by Biddles Ltd, Guildford

A catalogue record for this book
is available from The British Library

ISBN 1 870098 59 5

CONTENTS

PREFACE

✧

The papers in this volume come from participants at two recent conferences. The first was held at Biskops-Arnö, Sweden, and was jointly organised by the International Society for Ecology and Culture (ISEC) and Friends of the Earth, Sweden. The second, which took place in Leh, Ladakh, India, was organised by ISEC and its sister organisation, the Ladakh Ecological Development Group. These conferences sought to bring together thinkers and activists from around the world to discuss the root causes of the environmental and social crises facing the planet—in both 'developed' and 'developing' societies alike—and to explore some of the principles upon which to build a new model of development.

The papers cover a broad spectrum, from the very global to the very local. Some are theoretical works, others more based on personal experience; some were written specifically for this collection, others were excerpted from longer, previously published pieces, others still take the form of statements of principle.

Despite the variety of styles and content, the essential message is very clear: namely, that the process of development—whether in the North or the South—must change direction quite radically if we are to avoid ever more serious social and ecological problems. 'More of the same' can only be a recipe for disaster.

If there is one word which sums up the perspective which is now required, it is *diversity*. The notion of 'progress'—the dominant ideology of our age—implies the imposition of a single industrial model everywhere on Earth. What is needed instead is a model of development based on the capacity of people around the world to find ways of living based on their own cultural traditions and the full use of local resources, knowledge and skills. This fundamental principle is one which urgently needs to be recognised in the current debate on environment and development.

Editors' note: Some of the pieces which follow were written by people whose first language is not English. Although we have made changes in these papers when it was needed for the sake of clarity, we did not think it appropriate to 'anglicise' them completely.

I

Introduction

THE FUTURE OF PROGRESS

Helena Norberg-Hodge & Peter Goering

✧

From poverty and environmental degradation to overpopulation, ethnic friction and staggering international debt, the problems of the 'developing' nations of the South are all too familiar. Equally well-known are the crises facing the industrialised North: resource depletion, pollution, unemployment, crime, homelessness.

To cure these ills, most experts recommend the same industrial remedy, comprising equal parts economic growth (which today takes the form of 'free trade') and technological advance. Called 'progress' in the North and 'development' in the South, this regimen is commonly believed to offer long-term health for the Earth and prosperity for all its inhabitants.

The consensus of the authors in this volume is very different. From our perspective, it is clear that the industrial development model—far from offering solutions—is in fact a fundamental *cause* of present problems.

The many threads that make up this industrial model are closely interwoven, and the resulting tapestry is one of great complexity. Industrial society cannot be understood in terms of direct cause-and-effect relationships; it instead needs to be seen as a *system*, whose components are interrelated.

The economic paradigm

According to modern economics, a continuous increase in economic output is necessary, both to increase prosperity and to solve environmental and social problems. This belief, in fact, underlies the policies of *every* government, regardless of their position on the political spectrum. A narrowly defined criterion of economic efficiency is used to plan and administer economies, and factors that can be reduced to monetary value are given primary importance. Production choices are dictated by those who wield power in the money economy.

The economic growth imperative compels businesses to constantly grow, to find new markets, resources, and areas of life to colonise. Products are made to wear out sooner than necessary. Marketing professionals use whatever means are available, including the creation of new 'needs', to stimulate consumer spending.

The natural world is largely absent from the economic models used by development planners. There is an implied assumption that the Earth has an infinite capacity to supply the resources necessary for production, and to absorb the resulting wastes. In the real world, however, this is not true. It has now become clear that industrial society is seriously overburdening the biosphere, with potentially catastrophic consequences. Industrial society is in effect borrowing from future generations, which will inherit a depleted and degraded Earth.

The environment is not all that is left out of the models of conventional economists. Within the economic paradigm, attention is focused on those areas of life which can be most readily quantified. Values that help define human welfare—including happiness, fulfillment, morality and aesthetics—are essentially ignored.

The biases inherent in this way of thinking have a profound effect on development plans throughout the world. Purely economic measurements, for example, rate traditional subsistence societies as the poorest of the poor. Thus the UN lists Bhutan as one of the world's most impoverished countries, even though almost all of its people have adequate food, clothing and shelter, as well as sophisticated works of art and music—and more time for families and friends than most Westerners. According to modern economics, these nonmonetary measures of well-being are as nothing. What matters is GDP and per capita income, and on that count the Bhutanese are deemed to be no different from homeless people on urban streets.

Science and technology

The economic paradigm goes hand in hand with modern science and technology; together they form the driving force behind industrial society. Science gains its understanding of the world largely by isolating and studying small pieces out of the interconnected continuum of nature. This approach has had undeniable success, and modern technology is indeed able to manipulate the world to an almost unimaginable extent. However, the ability of scientists to predict the consequences of their actions is limited to narrow parameters implicit in the scientific method. Scientific models are most successful when dealing with the relatively simple and the short term; when it comes to the infinite complexity and long-term time frame of social systems or ecosystems, the limitations of science are particularly evident. Thus, while scientists today are feverishly racing to advance and implement the new biotechnologies, the long-term social and ecological implications of these 'scientific breakthroughs' cannot be revealed through scientific inquiry.

Given these fundamental shortcomings, the pre-eminent status of science today is profoundly disturbing. Science has come to dominate all other systems of knowledge. Traditions of non-Western cultures and the experience and intuition of individuals are accepted only to the extent that they can be verified by scientific observation. Meanwhile, the focus of scientific inquiry is getting narrower by the year, while its manipulations of the natural world deepen.

Infrastructure development and centralisation

Continuous economic growth and technological advance feed one another. Their dynamic interaction inevitably leads to large centralised organisations and requires a constantly expanding infrastructure: transportation and communication networks, centralised energy installations, large-scale irrigation schemes and Western-style educational facilities.

In the South, infrastructure projects are critical to the modernisation process. They are usually initiated by 'aid' programmes and funded by loans from international financial institutions such as the World Bank. A large portion of the overwhelming debts accumulated by Third World governments in the last three decades was incurred for infrastructure development—almost all of which is designed to serve the needs of international trade and the requirements of large urban centres. Very few resources go toward improving infrastructures that could enhance smaller communities or local exchange.

The failure of conventional economics to adequately account for environmental and social costs results in the perception that these larger production units offer 'economies of scale' and are more 'efficient' than small-scale decentralised structures. Similarly ignored are the massive subsidies for infrastructure development, which give large-scale centralised enterprises an unfair advantage over smaller, local units of production.

Centralisation and economic growth operate in a mutually reinforcing cycle: bigger and more profitable companies have more power to sway markets in their favour, better access to investment capital and information, and are better equipped for research and development. These advantages in turn help them to grow still bigger and to force out smaller competitors.

Urbanisation

The expansion of the industrial model through development thus promotes the breakdown of smaller-scale local economies. In both North and South, today's macro-economy forces people away from small towns and rural areas into ever larger urban conglomerates in search of paid employment.

Rural inhabitants find themselves on the periphery, reduced to resource suppliers for the cities where political and cultural life is centered. As rural areas decline in population and political importance, people's connection with the land is lost.

When individuals and communities control their own resources, they tend to behave responsibly: it is they and their families who directly benefit (or suffer) from the use (or misuse) of the environment. Through traditional practices, social stability and ecological balance are maintained. But in the centralised macro-economy, people are often far removed from the environmental impacts of heightened economic activity. Traditional behaviour patterns that once led to ecological harmony are abandoned, and many new problems—including rapid population growth—suddenly appear.

Social and environmental problems in rapidly growing cities are even more pronounced than those in rural areas. The sheer quantities of resources that are concentrated in urban centres overwhelm the ability of local ecosystems to absorb the impact of human activities. Natural cycles are disrupted; since people are separated from the consequences of their actions, even unmistakable signs of decline and breakdown can go unheeded.

People in today's urban conglomerates no longer need to depend on their neighbours. The interdependent communities that were an integral part of traditional life are transformed, in the modern setting, into collections of competing individuals. People are left insecure, alienated and isolated. This contributes directly to the problems facing modern urban populations around the world: crime and violence, alcohol and drug addiction, the abuse of women, children and the aged.

Ethnic strife

Many people in the 'developing' parts of the world recognise that modernisation is exacerbating ethnic rivalry, but tend to think of this as the necessary price of 'progress'. They believe that only through the creation of an entirely homogeneous and secular society can these rivalries be eliminated. Westerners, on the other hand, often assume that ethnic and religious strife is increasing because modern democracy liberates people, allowing old prejudices and hatred to surface. If there was peace earlier, they assume it was only because conflicts were repressed by authoritarian regimes.

The truth is very different. Through development, diverse people—often from very different cultures even within a single country—are pulled from rural areas into large urban centres, where community ties are broken and job opportunities are scarce. Young men who were once part of a social structure with a place for everyone must now fight for their survival, competing with others for jobs that are few and far between. In this artificially-created situation, any religious or ethnic differences are inevitably exaggerated and distorted. The situation is further aggravated by the fact that the ruling government usually favours its own kind, while members of other ethnic and religious groups are subject to discrimination.

On a deeper level, the modern development model gives rise to hostility by putting pressure on diverse people to conform to a Western model whose standards are impossible to meet. Most people in the developing world cannot be blue-eyed and blonde and live in two-car families. Yet—through the media and advertising—this is the image that is held up as the ideal. To strive for such an ideal is to reject one's own identity. The resulting alienation can give rise to resentment and anger, and can ultimately lead to violence and fundamentalism. In this book, Nsekyue Bizimana points out how the imitation of Western ways in his native Rwanda led to a breakdown of community and a rise in tribalism—precursors to the incredibly violent upheavals that have recently decimated the country.

Even people in the industrialised world are victimised by stereotyped media images, but in the Third World, where the gulf between reality and the Western ideal is so much wider, the sense of desperation is that much more acute.

The loss of self-sufficiency and sustainability
Since indigenous cultures were in tune with the specific resources and limitations of their local environment, the Earth's cultural diversity reflected its biological and geographic diversity. The people of high-desert mountain areas, for example, evolved cultures far different from those found on tropical islands. But industrial culture is based on technology and international finance, not the natural world, and its spread has had the effect of erasing diversity and leveling cultures. Today, virtually identical modern clothes, food and architecture can be found in every modernised city, regardless of the climate, local resources or cultural history.

In rural areas affected by development, even agricultural production is based on a single Western model. Under pressure from the industrial nations, the 'developing' countries have steadily abandoned their indigenous methods of food production and their basic-needs industries. Resources are increasingly exported to supply the demand for luxury items in the North, while Northern manufactured goods are imported back to the source country, where they replace locally-supplied goods. Farmers who once grew a variety of crops and kept a few animals to provide for themselves, now often grow a single cash crop for distant markets, and their livelihood can be jeopardised by disruptions in oil supplies or fluctuations in international commodity markets.

Previously sustainable ways of life are often supplanted by imported techniques that disregard the long-term impact on people and ecosystems. For instance, Martin Khor notes that in many 'poor' nations fish has traditionally been a cheap and plentiful source of animal protein. While traditional fishery methods served to maintain stocks and protect breeding beds, the new high-tech fishing industry has introduced methods which deplete fish stocks faster than they can be replenished. As a result, livelihoods in small-scale fishing communities in the Third World is threatened—and the cost of fish has gone up, so poor people can no longer afford it.

Virtually overnight, development can bring environmental degradation to regions where people have lived sustainably for millennia. According to Filipina del Rosario-Santos, the Philippine forests will last only 35 more years if present harvest rates continue. This decimation of rainforest results in severe topsoil erosion, loss of biodiversity, climatic changes, and the displacement of the human communities which depend directly on the forest for their livelihood.

Population growth
It is generally forgotten that populations in traditional societies—including Old Europe—remained relatively stable for centuries. Edward Goldsmith describes some of the strategies and taboos

traditional societies employed to control population. Moreover, these cultures had a direct relationship with the resources of the land around them, and were thus better able to match their numbers to the land's carrying capacity. It was only when these traditional patterns were disrupted that populations began to explode.

The conventional wisdom is that the only way to reverse this explosion is through more economic development, so that the South can come to enjoy the same material standard of living as the North. On the face of it, this is a perfectly reasonable argument, since even the most cursory look at the global demographic map will show that population growth is higher in the less industrialised parts of the world and lower in the more industrialised parts.

However, the mathematics of the argument simply do not work. The North's prosperity is only possible because its inhabitants, one-fifth of the world total, consume roughly four-fifths of the earth's resources—an imbalance that continues to grow. What's more, the biosphere is already under serious threat from current levels of human activity: it is inconceivable to imagine that the Earth could sustainably absorb the 16-fold increase in economic activity that would be needed for the South's consumption level to equal that of the North.

A related notion is that limiting population growth in the South will require more and 'better' education for women, leading to job opportunities outside the home and family. This ignores the fact that the education prescribed is specialised training for jobs in a global economy, in which there simply are not enough jobs to go around. Western-style education for women is merely another aspect of a development process that hastens the breakdown of more diversified locally-based economies and accelerates urbanisation.

The industrial model is simply not a pattern which is replicable around the world. Even if it were technically feasible, it is highly questionable whether a model which leads inevitably to such serious social and environmental problems—family and community breakdown, poisoned soils and water, a destabilised climate—should be imposed everywhere.

Cultural breakdown

Few people question the modern education system, which has been promoted throughout the developing world. But while traditional education teaches people how to live and work within a particular ecosystem, Western-style schools train people to become narrow specialists in a Westernised urban environment. This education cuts children off from the accumulated knowledge about local resources that has been passed down from generation to generation. In so doing, modern education contributes to both the breakdown of local culture and the loss of self-esteem.

Schoolchildren in many non-Western cultures rarely learn anything directly relevant to their environment and resources. Instead they receive a poor version of a New Yorker's education—including learning Wordsworth and English translations of *The Iliad*, as is the case in Ladakh. Evelyne Hong describes a similar situation in East Sarawak, and the cultural breakdown that ensues.

Cultural diversity is also undermined by development and modernisation through the creation of artificial 'needs'. The image of the industrial 'good life' is carried around the planet by the media, advertising and education. It is an image that in reality exists nowhere, but it makes the daily existence of almost everyone seem boring, inadequate and poor by contrast. In diverse cultures around the world, people have changed from being proud and self-reliant into a people who are ashamed not to have 'modern' conveniences, and who now 'need' not only imported cement and packaged foods, but also mirrored sunglasses, designer blue jeans, and Rambo T-shirts.

The craving for luxury items in both the First and Third Worlds exacerbates the depletion of resources, particularly from tropical areas. The élites in the Third World tend to copy the lifestyle of their Northern counterparts, even buying expensive imported goods manufactured from resources taken from their own countries—thus further impoverishing their own populations. S.M. Idris points out that billions of Third World dollars are spent on imported automobiles and in building roads and bridges, while most public transportation systems in the Third World are quite inadequate.

The myth of progress

The concept of 'progress' lies at the heart of the ideology of industrial society. Modern technology and economic growth, it is believed, will ensure ever-increasing prosperity. It is assumed that all societies will eventually follow the same path as the most 'advanced' industrial societies. This linear conception of development inherently places some societies 'ahead' and some 'behind'. The label of 'backward' or 'underdeveloped' creates tremendous psychological pressure to embrace what is perceived to be more modern or advanced. Since the goal is continually changing and perpetually out of reach of all except the most privileged minority, the system inevitably creates insecurity and anxiety.

A close look at the limits to resources and at current social and environmental crises shows that the belief in progress is a cruel myth. Leaving aside the question of whether 'modern' lifestyles are really desirable, it is clear that there are simply not enough resources for the entire world to duplicate the consumption patterns of the richest countries, nor indeed are there sufficient resources for the rich countries to maintain their present level of consumption. Technological innovation is offered as the magical way around these resource limits, but this too is an empty promise. It obscures the fact that each new technology brings about unpredictable social and ecological consequences, and that many of our most pressing problems are the *result* of technologies previously heralded as godsends. Human beings are far from able to understand and control the natural world, and are ultimately dependent on natural processes—not technology—for the necessities of life.

Conventional Solutions

While many international institutions and national governments are beginning to recognise some of the problems inherent in the current development model, the solutions currently being proposed in most fora fail to address their root causes. For example, *Our Common Future* (the report of the World Commission on Environment and Development, often referred to as 'The Brundtland Report'), calls for renewed commitment to economic growth and high technology—albeit with environmental reforms—in order to generate the wealth necessary to alleviate poverty and clean up the environment. And most development organisations view the transfer of power to supranational institutions (rather than the re-empowering of local, self-reliant communities) as a positive trend. Similarly, they support the establishment of Western-style schools and universities—an educational system that emphasises specialisation and leads to the loss of local knowledge.

Free trade

'Free trade' is seen by government leaders as a panacea for world economic problems. However, the reality is that free trade policies work against the interests of the vast majority of producers and consumers—in both North and South—by systematically dismantling locally-based economies.

Free trade agreements aim at allowing transnational corporations to establish markets and subsidiaries anywhere, with as few restrictions as possible. Locating operations in the South allows corporations to take advantage of lower labour costs and less strict environmental regulations. Moving factories and industrial agriculture operations to the developing world not only exacerbates unemployment in the North, but also displaces Third World people from their traditional occupations, particularly in agriculture. Rural areas will be even more rapidly depleted as people are drawn into already overwhelmed urban areas.

Free trade also challenges the regulations established by Third World countries to protect their resources. Indonesia, for instance, recently attempted to ban the export of Indonesian rattan, which would help protect a rapidly dwindling forest resource. The US and members of the EC immediately protested, saying that this was an unacceptable barrier to trade.

The opening up of every economy in the world will make it easier for corporations to market pesticides and pharmaceuticals in the South that are banned in the North. The Third World has already become a destination point for hazardous waste produced in the North, and efforts to limit this practice are undermined by free trade policies.

Martin Khor draws attention to the Uruguay Round of GATT, which, he says, will usher in still greater deregulation of Northern industry, particularly in the area of services. Proposals sponsored by the US and other developed nations will allow the unregulated influx of Northern service industries into the South. This will remove banking, health care, media, and communications from local control and put it in the hands of multinational corporations.

As has been pointed out, constant exposure to Western media and advertising images is one of the factors undermining diverse cultures and giving rise to social and psychological breakdown. 'Free trade' treaties will make it all but impossible to control or limit the influx of these culturally destructive images.

Industrial agriculture

No single area of life has been more profoundly affected by the process of industrial development than agriculture. And since agriculture plays such a central role in non-industrialised societies, the impact of the changes which development has brought about is particularly widespread.

Agricultural experts—focussing on narrow parameters in a simplified model—promote Western methods of agriculture in order to increase production. The export of this agricultural technology—including farm machinery, hybrid seeds, chemical fertilisers and pesticides—opens new markets for industries in the North, a necessary ingredient for its continued economic growth. The South is promised that their increased dependence on Northern inputs will be more than offset by the wealth created from higher agricultural productivity.

However, the models employed by the agricultural technologists do not match the complexities of the natural world. Unlike the varied strains of a given crop grown traditionally, new strains of genetically identical hybrids are not the result of slow adaptation and evolution in the host environment. The new strains are more vulnerable to the exigencies of weather and pests, and increasing amounts of fertilisers and pesticides are required to maintain yields. Rather than the promised prosperity, many Third World countries instead find themselves with degraded agricultural land, continued reliance on Western agricultural technology, and a less-than-adequate food supply.

The new biotechnologies represent a further intensification of this process. Much of the genetic information used in genetic engineering originally comes from wild or domesticated plant varieties found in the South. These expropriated resources are then to be sold back to the South as part of 'patented' forms of life, which require farmers to purchase new seeds, fertilisers and herbicides year after year. Among other things, agro-chemical corporations are working to genetically engineer crops to be resistant to the herbicides the corporations themselves produce and sell. The long-term ecological risks of this genetic tampering are unknown, but the economic result will be further dependence on transnational corporations.

Strategies for Real Change

Strategies for change fall into two distinct categories: those that counter destructive trends, and those that help foster more positive alternatives. Each of these requires different approaches. 'Counter-development'—opposing the further extension of industrial monoculture—requires massive and rapid efforts, both locally and internationally. On the other hand, implementing more sustainable alternatives is work that needs to proceed slowly and carefully, and will vary significantly from place to place.

Many regions of the South, where the way of life is still based on traditional and ecologically-sensitive patterns, contain the seed for their own sustainable future. Recreating strong local economies and vibrant human-scale communities is a much bigger challenge in the highly urbanised North. Perhaps it will be the South, where village settlements are still the norm, which will lead the world toward saner ways of living.

Challenging the techno-economic model
A vital part of a counter-development strategy is to publicise the

facts about the environmental degradation and social problems of the North. It is especially important for this information to reach the South, where the problems of the 'modern' world are not widely known. Stephanie Mills describes some of the consequences of misguided development in the US, a story that needs to be heard more widely in the South.

The testimony of people from the South who have experienced the reality of life in the most advanced industrial societies can be one of the most powerful ways of correcting false impressions. Dr. Nsekuye Bizimana tells how his expectations of Western life far exceeded what he found when he actually went to live there, and how his experiences in the West enabled him to better understand the breakdown in his native Rwanda.

People in the North also need to know more about the reality of life in the South. Above all, perhaps, they need to understand what 'development' really means in practice; that far from raising Third World living standards, as is commonly supposed, development has tended overall to impoverish people by making them dependent—both physically and psychologically—on international economic and political forces far beyond their control.

The educated classes of the North who enjoy privilege in the current world order have a special responsibility for implementing change. They have many more resources at their disposal and much greater influence than the marginalised classes of North and South, who are often struggling just to survive. The Northern democracies also tend to be more responsive to citizen pressure than the often authoritarian regimes of the South.

There are many creative ways of building solidarity between the populations of North and South while raising awareness of the inter-connections between the two groups. Silvia Ribeiro and Birgitta Wrenfelt describe how Future Earth Sweden links grassroots groups working on similar socio-ecological issues. The network creates and supports projects based on local self-reliance, promotes information exchange between like-minded groups, and provides training on issues of appropriate technology and alternative development.

Control over information is now concentrated in the hands

of governments and large corporations, which are biased toward the perpetuation of the current system. Alternative viewpoints and analyses are filtered out, personal thoughts and experiences are devalued, and many important events and issues are never discussed. Although it would be costly and centrally organised, a large-scale alternative media campaign is nonetheless consistent with a philosophy of small-scale grassroots organising. A massive publicity campaign may be the only way to stop the development juggernaut and to create the space for alternatives to evolve.

The policies of a few international institutions, such as the World Bank, the International Monetary Fund (IMF), the General Agreement on Tariffs and Trade (GATT), the Food and Agriculture Organisation (FAO), and the World Trade Organisation (WTO), affect the lives of people all over the world. Yet decision-making in these institutions is not democratic, and there is no real forum for public input. Campaigns to change these institutions and to force them to live up to their official statements about environmental sustainability and social justice are a crucial part of the process.

Existing examples include the network of activists organised by *The Ecologist* to expose the unsustainable policies of FAO; the concerted efforts by dozens of grassroots organisations which succeeded in persuading the World Bank to abandon its support for the Narmada dam in India; and the '50 Years is Enough' campaign against the Bank and the IMF, which involved activist organisations from around the world.

The work of Vandana Shiva and colleagues on biodiversity is another example. International institutions such as GATT and the FAO are currently being pressured by multinational corporations to recognise genetic resources as a 'universal heritage' (rather than a locally-controlled resource) to ensure the North's free access to raw materials. In opposition to this, Shiva and associates have developed a 'People's Biodiversity Conservation Action Plan' that would recognise local stewardship of genetic resources, monitor and analyse the activities of international agencies in this area, and encourage protection and regeneration of biodiversity in farming and forest use.

Many of the contributors to this volume are involved in other programmes aimed at resisting the process of monoculturisation:

- Vandana Shiva is closely involved in the Chipko movement, which was founded in the 1970s in response to the massive deforestation taking place in the Himalaya. Over the last two decades, people from a number of small mountain communities have successfully slowed the destruction of several Himalayan forests.

- Members of the Norwegian Ecopolitical Community against the European Community (NØFF) successfully opposed Norway's entrance into the EC, on the grounds that it would lead to non-sustainable development patterns and the erosion of Norway's unique culture.

- Martin Khor and Mohamed Idris have been active in alerting people to the implications of the Uruguay Round of the General Agreement on Tariffs and Trade (GATT), the 'free trade' agreement which aims at making every corner of the planet a source of profit for transnational corporations. In the process, they point out, it will increase the disparities in wealth between North and South and hasten the depletion of natural resources worldwide.

It is absolutely critical that the movement to resist the conventional development model be as broad-based as possible. Single-issue pressure groups certainly have an important role to play, but they need to be based on an understanding of 'the bigger picture'. While it is of course extremely important, for instance, to block the construction of particular dams in particular parts of the world, or to prevent the destruction of particular areas of rainforest, it is equally important to lobby for funds for public education to show why the construction of large dams or the cutting down of rainforests is *in general* unsustainable. By linking individual actions to a larger vision, the counter-development message will have far greater impact.

Equally, we need to encourage individuals and groups with broadly similar goals to collaborate and speak with a common voice, thus greatly increasing their impact. Not least, we need to bring together people working on environmental issues with those campaigning for social justice—thereby forging, in the words of Nicholas Hildyard, the new path of 'liberation ecology'.

Fostering small-scale alternatives

Countering further colonisation by the industrial monoculture is only half of the strategy needed if we are to move towards more sustainable ways of life. Equally important is the creation of many and diverse alternatives.

A smaller scale is critical if we are to adapt to natural systems rather than attempting to manipulate nature—and human needs—to fit techno-economic structures. Intrinsic to this approach is a fundamental humility concerning humanity's ability to control and manipulate the natural world. Instead of technological hubris we need to encourage a spiritual communion with all life. The psychological and spiritual vacuum we now try to fill by grasping for material possessions can be eliminated through a deeper connection with nature, stronger community ties, and the integration of work and leisure.

Many of the contributors to this volume are active in creating and nurturing such initiatives:

• In Sweden, Erni and Ola Friholt are trying to reinstate a sustainable way of life in Stocken, a small fishing village. They ride bicycles instead of driving a car, have reduced their personal consumption to 1950s levels, and run a café using mainly locally produced raw materials. They are working to create a locally-based alternative, by consciously detaching themselves from the global economic system.

• Märta Fritz describes the growing 'eco-village' movement in Sweden. These communities favour low water and energy consumption, maximum levels of recycling, and the use of nontoxic building materials. And in stark contrast to the trends in mainstream culture (63% of Stockholm's dwellings are one-person homes), these villages seek to nurture a real sense of community.

- Stephanie Mills is working with others in the broad-based 'bioregional movement' to restore the ecological stability and diversity of the many bioregions in the United States. Part of her work is in the restoration of ecosystems which have been levelled by monoculturisation.

Appropriate technology

Many groups are working in the South to promote small-scale decentralised technologies that use local resources. The Ladakh Project and the indigenous Ladakh Ecological Development Group (LEDeG) have developed a number of effective technologies using solar energy, and have expanded on the traditional use of water power. S.M. Idris, Martin Khor and the Consumers' Association of Penang are also involved in supporting appropriate technology in both rural and urban areas of Malaysia. In Rajastan, India, Aman Singh is working with local villagers to resurrect successful small-scale technologies that have been swept aside by development and dismissed by scientific 'experts'.

At the governmental level, however, appropriate technology receives almost no support. Not one of the 'developing' nations has made solar energy the focus of its energy policy, despite the abundance of this free energy source in many areas of the South.

In most industrialised countries these technologies cannot compete with centralised energy sources like fossil fuels and nuclear power, which have benefited from decades of subsidies in the form of infrastructure investment and ignored environmental costs. Nowhere is this more true than in the United States, where direct and hidden subsidies for centralised energy technologies run to well over $50 billion a year. If there is to be any hope of restoring social and environmental balance, this situation must be gradually reversed, giving support instead to efforts to promote a decentralised renewable energy infrastructure.

Many solar and small-scale hydro and wind power technologies lend themselves well to decentralised settlement patterns in both the North and South. The local or regional provision of basic needs such as food and energy would eliminate a large proportion of current transport costs. The savings on the construction and maintenance of infrastructure could be used to promote local development.

Alternatives in education

While education in nature-based societies emphasises the interconnections among diverse elements of the surrounding world, Western education compartmentalises fields of learning, leading to a narrowly-focused over-specialisation. Some encouraging trends have, however, begun to appear. Many respected institutions of higher learning in the West are implicitly acknowledging the limitations of specialised education, and are now using an interdisciplinary approach in some fields. For example, the Human Ecology Group at the University of Edinburgh, and the Energy and Resources Group at the University of California at Berkeley reduce the boundaries between such previously separate disciplines as the social and 'hard' sciences, leading to a more complete picture of the interdependence and complexity of the real world.

Other institutions have even more radically transformed their approach to education. A prime example is Schumacher College in England, which challenges the quantifying and mechanistic view of the world, and places emphasis on more holistic and spiritual learning. In the US, the California Institute of Integral Studies offers degree programmes in fields that synthesise East/West thought and values, while Antioch University in Seattle offers a programme which approaches diverse fields from a holistic, 'systems' perspective.

Synthesising traditional and post-industrial values

In the industrialised countries of the North, people have been dependent on the industrial system for so long that most have lost touch with past traditions. However, the seeds of a new way of life are beginning to be planted.

In fields as diverse as hospice care for the dying and mediation as away of settling disputes, striking parallels are emerging between the most ancient and the most modern. Increasing numbers of people are choosing to eat wholefoods that are grown naturally, and are using age-old nature remedies for their health problems. There is a reawakened interest in storytelling, a renewed appreciation for physical work, and a growing preference for natural materials in clothing and construction. As in traditional, pre-industrial cultures, this new post-industrial culture lays great stress on the importance of community—on a sense of cooperation and interdependence with others and with local natural resources. Like traditional cultures, the new movement greatly values spiritual wisdom, a deep sense of our connection to the rest of life—of living in the here and now.

Rural populations of the South have a great deal to offer the North in all of these areas, as well as in such specific ways as attitudes towards time and property; social mechanisms of reciprocity and cooperation; the maintenance of communities that protect the interests of the young and the old; knowledge and technologies that have grown out of an intimate familiarity with a specific place; and the art of living within the limits of the natural world.

Even though we in the North can't 'go back' to living like the indigenous peoples of the Amazon rainforest or the Tibetan Plateau, we *can* learn from such societies. These cultures are, in fact, the only time-tested models of truly sustainable practices on Earth, yet they are rapidly disappearing in the name of 'progress.'

II

Environment
and
Development

DEVELOPMENT, TRADE AND THE ENVIRONMENT:
A Third World Perspective

Martin Khor

❖

Around the world, governments and citizens' groups alike are beginning to focus on a wide range of issues relating to the world environmental crisis and the survival of the Earth and humanity.

It is becoming increasingly clear, however, that the technological fix-it solutions favoured by the West are not enough. Indeed, they are not even of central importance to the solution of the environmental crisis. Instead, social, developmental and political issues need to take centre stage. For the main issues emerging in the ecological debate revolve around the control, distribution and use (or abuse) of the world's increasingly scarce natural resources.

Many people from the Third World are complaining that the agenda for reform has been too dominated by Northern interests, that the North is only interested in the physical and environmental aspects of the crisis, while neglecting the Third World's need for development. It is feared that focusing on technical aspects of ecological problems without putting them in the context of unequal North-South economic relations will lead to another form of domination over the Third World. The South could be asked to stop or slow down its development while the North, already enjoying high living standards, would continue its way of life with only minimal inconveniences from technological adjustment.

Threats to the well-being of the peoples and environments of the Third World will only worsen as a result of proposals contained in the Uruguay Round of GATT (General Agreement

33

on Tariffs and Trade). This trade agreement allows transnational companies major influence in many crucial areas previously denied them in Third World countries. It is very likely that the environmentally and socially exploitative effects of this GATT agreement will have a far greater impact than any constructive proposals put forward at the United Nations Conference on Environment and Development (UNCED), held in Brazil in 1992. The entire UNCED process, in fact, was seen by many as a cosmetic exercise set up by the developed countries to give the impression that they are seriously concerned about Third World problems.

The UNCED conference, and particularly the Global Forum, nevertheless provided a much needed opportunity for the Third World to present its analysis of the roots of the global environmental crisis as it manifests itself in the Third World, and for people there to advocate a different form of society and a different type of 'development'. It also provided a chance for the Third World to point out that its environmental problems must be seen not only from the level of the community or even the nation, but as an integral component of a global system of industrialism. This system shapes the way and the rate in which resources are used, and determines not only how environments are recreated or disrupted, but also how people's very lives are affected.

The following analysis focuses on Third World environmental problems as products of the importation of the Western system, on the exploitative transfer of resources from South to North, and on the negative effects on the Third World of the Uruguay Round of GATT. It also suggests steps which might be taken to bring a new environmental order into being.

The Conquest of the Third World

In Third World societies before colonial rule and the infusion of Western systems, people lived in relatively self-sufficient communities. For food they planted staple crops like rice or barley, tended animals, and fished and hunted; they supplied housing, clothing and other needs through small-scale industries which made use of local resources and indigenous skills. The mode of production and style of life were in harmony with the natural environment.

Colonial political rule—accompanied by the imposition of new economic systems, new crops, industrial exploitation of minerals and participation in the world market (with Third World resources being exported and Western industrial products being imported)—changed the social and economic structures of Third World societies. The new structures, consumption styles and technological systems had become so in-built that after the attainment of political independence, the importation of Western values, products, technologies and capital not only continued but expanded. Third World countries became more and more tied into the world trading, financial and investment system. Transnational corporations were in the forefront, setting up trading and production bases in Third World countries and selling products and technologies to them. They were aided by an infrastructure of aid programmes funded by the rich-country governments, by multilateral institutions like the World Bank, and also by transnational banks, which loaned billions of dollars to finance expensive projects and highly capital-intensive imported technologies. They were also supported by foundations, research institutions and scientists in the rich countries, which sponsored and carried out research on new agricultural technologies which would 'modernise' the Third World—in other words create the conditions whereby the Third World would have to depend on the transnational companies for technology and inputs.

In order to pay for the import of modern technology and inputs, Third World countries were forced to export even more goods (mainly natural resources like timber, oil and other minerals) and export crops which took up a larger and larger portion of total land area. Economically, financially and technologically, Third World countries were sucked deeper and deeper into what has proved to be the whirlpool of the world economic system. In the process, the Third World has lost or is in the process of losing its indigenous products, resources and skills. Our peoples are losing their capacity for self-reliance, their confidence, and in many cases the very resource base on which their survival depends. The world economic and technological systems are themselves facing crises. The Third World is now hitched onto these systems over which they have very little control. The survival and viability of most Third World societies will thus be put to the test in the next few decades.

There are numerous examples of how the Western system has resulted in the degradation of the environment and the deterioration of human health in the Third World:

Import of hazardous technologies and products

Many transnational companies have shifted their production operations to the Third World, where safety regulations are either very lax or nonexistent. Many industries are also shifting their sales promotion and markets to the Third World, where they can sell products of lower quality or products which are so toxic they are not allowed to be used in the rich countries themselves.

As a result, many Third World countries are now exposed to extremely toxic or dangerous technologies which could potentially cause great harm. The Bhopal gas tragedy, in which 3,000 lives were lost and another 200,000 people suffered disabilities, is the worst single example of what can happen when a Western transnational company adopts industrial safety standards far below acceptable levels in its home country. There are hundreds of other substandard industrial plants sold to the Third World (the Bataan nuclear plant in the Philippines, for example) or shifted there by transnationals to escape health and pollution standards in their home countries.

Hazardous products are also being pushed on the Third World in increasing volume. Examples include pharmaceutical drugs; contraceptives; pesticides banned years ago in European and American countries or in Japan, but sold by companies of these same countries to the Third World; cigarettes containing far higher levels of tar and nicotine than in the rich countries; and, most recently, milk products contaminated with radioactivity from the Chernobyl nuclear disaster. The health effects on Third World peoples are horrendous. For example, it is estimated that

40,000 people in the Third World die from pesticide poisoning each year. Moreover, millions of babies are also estimated to have died of malnutrition or illness through taking diluted or contaminated commercial milk, after their mothers gave up breast-feeding on being persuaded that bottle milk was superior.

The hazardous technologies and products imported from the rich countries also often displace indigenous technologies and products which may be more appropriate to meet the production and consumption needs of the Third World. Labour-intensive technologies which are in harmony with the environment while providing employment for the community (for instance, traditional fishing methods) are replaced with capital-intensive modern technologies which are often ecologically destructive. More appropriate products or processes (for example, breast-feeding) are replaced by modern products which are thrust upon the people through high-powered advertising, sales promotion and pricing policy. The Third World is thus in the process of losing many of its indigenous skills, technologies and products which are unable to stand up to the onslaught of the modern system.

The 'Green Revolution'

The imposition and penetration of the Western system has also changed the face of Third World agriculture. In many Third World societies, a large part of the lands planted with traditional food crops was converted into cash crops for export under the plantation system. When export prices are high, the incomes of export-crop farmers may be high; but when prices fall (as in the present period) these same farmers may be unable to buy enough food with their incomes, and many agricultural workers may lose their jobs.

In those areas where farmers still cultivate food crops, the so-called Green Revolution has had a tremendous impact. The Green Revolution is a package programme which aims at increasing production through the introduction of high-yielding seed varieties, high doses of chemical fertilisers and pesticides, agricultural machinery and irrigation. In many areas where this modern-package 'revolution' was implemented there was an initial rise in production because more than a single crop could be produced in a year. But the rise in farmers' incomes was soon reduced by the increasing costs of imported chemical inputs and machinery. The pesticides used also exacted a heavy toll in thousands of poisoning cases. It is now found that the high-yielding varieties are increasingly susceptible to pest attack as the pests become resistant to the pesticides. Yields in some areas have begun to drop. Meanwhile thousands of indigenous rice varieties that had withstood generations of pest attacks have been wiped out. Many of these seed varieties are now located only in research laboratories, most of which are controlled by rich countries and international agencies. Third World farmers and governments will increasingly be at the mercy of the transnational food companies and research institutions—which have collected and patented seeds and germ plasm originating from the Third World.

Biotechnology

Biotechnology is becoming an increasingly important scientific weapon by which the First World can increase its domination of the Third World. Biotechnology alters the structure of the DNA within organisms in order to bring about desirable characteristics in plants and animals. Although it is a relatively new field, biotechnology has already had severely detrimental impacts on Third World economies.

Examples of this include: genetically engineered fructose, which has captured over 10% of the world sugar market and has caused sugar prices to fall, throwing tens of thousands of sugar workers in the Third World out of work; 'natural' vanilla beans—produced in the biotech labs of a Texas firm—which have eliminated the incomes of 70,000 people growing vanilla in Madagascar; and a new industrial process for producing natural gum discovered by a New York company, which cost Sudan its export market for gum arabic.

It is now estimated that biotechnology can find substitutes for $14 billion worth of Third World commodities now exported to the rich countries annually. Such a displacement could dramatically reduce the Third World's income.

Modern fishing technologies

The Third World's fishery resources have also been adversely affected by the importation of Western high-tech fishing technologies. In many Third World societies, fish is the main source of animal protein, and fishing used to be a major economic activity. In traditional fishing, the nets and traps were simple, and ecological principles were adhered to: the mesh size of nets were large enough so as not to trap small fish, and since breeding grounds were not disturbed, fish stocks could multiply. Fishing required hard work and tremendous location-specific knowledge, passed on through generations. Boats and nets were usually made from local materials; fishing—including fish preservation, mending of nets, making of boats, and so on—involved the whole community.

Then modern trawl fishing was introduced, in many cases as a result of aid programmes provided by rich countries. (In Malaysia, for instance, trawl fishing was introduced through a German aid programme.) There was a massive increase in the number of trawlers, usually owned by non-fishing businessmen and operated by wage-earning crews. There was gross overfishing, and much of the fish caught by trawlers was not used for human consumption but was sold instead to factories as feedmeal for animals. The criterion in trawl fishing was 'maximum catch in the short run for maximum present revenue'. The mesh size was usually so small that even small fish could be netted and sold. Destructive gear was used which scraped the bottom of the seabed and disturbed breeding grounds. As a result, there has been a depletion of fishery resources, and the catch for both traditional and trawl fishermen has declined in many parts of the Third World.

Riverine fishery resources have also been hit hard by technologies imported from the North. Toxic effluents from industrial estates kill off fish life, as well as poisoning villagers' water supplies. In the rice ponds, where farmers previously caught freshwater fish to supplement their diet, the pesticides introduced through the Green Revolution have also killed off fish life. This has threatened the livelihood of millions of small-scale fishermen in the Third World, while reducing an important source of protein for the general population. In Malaysia, where fish used to be abundant and was considered a poor man's meat, marine seafood is now about the most expensive item on restaurant menus because of the depletion of marine life. For poor people, access to fish is now more and more restricted because of decreased quantity and escalating prices on the market.

Logging of tropical forests

Another Third World resource fast disappearing is the tropical forest. Traditionally the forests were inhabited by indigenous peoples practising swidden agriculture which—contrary to propaganda from the modern system—was an ecologically sound agricultural system causing minimal soil erosion in the hilly tropical terrain. This system has been threatened by massive logging activities as trees are chopped down by timber companies for export to the rich countries, or for conversion of primary forest to grazing land for cattle for the hamburger industry in the United States. Between 1900 and 1965, half the forest area in developing countries was cleared, and since 1965 the destruction of the forest has accelerated even more. Many millions of acres are destroyed or seriously degraded each year, and by the turn of the century little primary forest will be left.

The massive deforestation has multiple ecological and social consequences, including the loss of land rights and way of life (or even life itself) of millions of tribal peoples throughout the Third World; massive soil erosion due to the removal of tree cover, thus causing the loss of invaluable topsoil; much reduced intake of rainwater in catchment areas as the loss of tree cover increases water runoff to rivers; extensive flooding in rural and urban areas caused by excessive silting of river systems; and climatic changes due to increased carbon dioxide in the atmosphere caused by the loss of trees. Besides these effects, there is also the loss of biological diversity as plant and animal life disappears from these ancient forests.

Modern industrial plants and energy megaprojects

The introduction of Western consumer goods, industrial plants and energy megaprojects has also greatly contributed to the loss of well-being in the Third World.

The indigenous small-scale industries of the Third World used to produce simple goods which were required to satisfy the basic needs of the majority of people: processed food, household utensils, footwear and clothes, furniture, simple houses and so on. The technology to manufacture these goods was simple and labour-intensive. Many of these indigenous industries have been displaced by the entry of modern products which, when heavily promoted through advertising and salesmanship, become glamourised, thus rendering local products unglamorous and low in status by comparison. With modern products capturing high market shares, capital-intensive industries (usually foreign-owned) set up base in the Third World and displace traditional locally-owned industries.

But many Third World countries were not content simply with modern consumer-goods industries. They also ventured into very expensive heavy industries and industrial projects such as steel mills, cement plants, and large infrastructure projects such as long highways, big bridges and super-tall buildings. These are projects and infrastructure copied from the cities of the rich countries. The political leaders of the Third World feel their countries need to have these super projects in order to become 'developed'.

To cater to the huge industrial plants and infrastructure, huge amounts of energy are required. So began the megaprojects in the energy sector, in particular large hydroelectric dams and nuclear powerstations. Each such project has its problems. The huge dams require the flooding of large tracts of land, causing the loss of forest and disruption to the livelihood and way of life of the many thousands of people living there. The dams themselves do not have a long lifespan due to siltation, so they are usually not viable financially; in other words, their costs far outweigh their benefits. There are health effects, as ecological changes associated with dams spread schistosomiasis disease (carried by snails) as well as malaria and other ailments. There is of course also the possibility of a major tragedy should the dam burst.

In the case of nuclear power plants, there is the possibility that those plants sold to the Third World may not have the same standard of quality and safety as those installed in rich countries where there is stricter quality control and technical expertise. If a plant installed in a Third World country is found unsafe, the government has a dilemma: stop its operation and incur a huge loss, or continue using it but run the risk of a tragic accident. In the Philippines, Westinghouse Corporation built a nuclear plant for $2 billion but there are so many doubts about its safety that the Philippine government has decided to 'mothball' it. Even if a nuclear plant is initially declared safe enough to start operating, there is no guarantee against a serious accident. There is also the ongoing problem of disposing of its radioactive waste.

In all these huge industrial, infrastructural and energy projects, a lot of money is involved, often running to hundreds of millions or billions of US dollars. These projects are invariably marketed by transnational companies who stand to gain huge sums in sales and profits per approved project. Financing is arranged for by the World Bank, by transnational commercial banks, or by rich-country governments, usually under aid programmes. Such projects are usually inappropriate for genuine development, since they end up underutilised, grossly inefficient or too dangerous to use. Absorbing so much investment funding, they deprive communities of much-needed financing for genuine development projects, and moreover lead Third World nations into the external debt trap. Particularly in the case of dams, they cause widespread disruption and displacement of poor communities, especially indigenous peoples, who by the tens of thousands have to be 'resettled' as their forests and lands are flooded out.

The Drain of Resources from South to North

The above examples illustrate how the penetration of industrial-nation technologies and projects can result in disastrous consequences for Third World environments and peoples. But this is only part of a long historical process, which now has unfortunately intensified, in which environmental and other resources have been transferred from poor to rich countries. The control of both technological capacity and the systems of world trade and finance has enabled the industrial countries to suck out forest, mineral and metal resources from the Third World, and to make use of its land and labour to produce the raw materials that feed into the machinery of industrialism. The rich nations—with around a fifth of world population—use up four fifths of world resources, a large portion of which is used for making luxury products. The Third World gets to use only 20% of resources. Since incomes are also unequally distributed within Third World nations, a large part of these resources are also used up in making or importing the same high-tech products as are enjoyed in the rich nations, and in importing capital-intensive technologies to produce these élite consumer goods. Thus, only a small portion of world resources flows towards meeting the basic needs of the poor majority in the Third World.

This is the ultimate environmental and social tragedy of our age: the scientific knowledge that could be properly used to provide for every human being's physical needs is being applied instead through industrial technology to take away resources from the Third World, largely for the production of superfluous goods. Meanwhile, the majority of Third World peoples sink deeper into the margins of survival.

Worse yet, the very processes of extracting Third World resources result in environmental disasters—deforestation, soil erosion, desertification, pollution of water supplies—as well as the horrible human toll in poisoning from toxic substances and in industrial accidents. The resource base on which communities have traditionally relied for both production and home needs has been rapidly eroded. Soils required for food production become infertile, forests which are home and life for indigenous peoples are logged or flooded out, water from the rivers and wells are clogged up with silt from deforestation or with toxic industrial effluents, and twigs and branches from trees used as a renewable energy supply disappear as the whole forest is axed by timbermen.

The transfer of resources from South to North takes place through many channels. First, there is the transfer of physical resources, as logs and metals and oil are shipped from South to North. For example, the developed countries produce and keep 80% of the world's industrial wood but also import much of the rest of the world's timber harvest as well; only 20% of the world's industrial wood comes from tropical forests, but more than half of that is exported to the richest nations. Most of it is used for furniture, high-class joinery, housing, packing material, even matchsticks. The timber that is exported to rich countries is lost to Third World peoples who now find it difficult to get wood for essential use in making houses, furniture and boats.

Secondly, there is a transfer of financial resources in that the prices of Third World commodities (obtained from their environmental resources) are low and declining even more. Between $60 and $100 billion were lost to Third World countries per year in 1985 and 1986 alone due to the fall in commodity prices. In human terms, this means drastic cuts in living standards, massive retrenchments of workers, and big reductions in government budgets in many Third World countries.

Thirdly, many of the 'development projects' which lead to the loss of resources are financed by foreign loans. It is rare for these projects to generate sufficient returns to enable repayment of the debts. In the end, a country loses valuable foreign exchange and income in debt repayment. Moreover the Third World is rendered ever more dependent on the technologies and products of the industrial world, for which substantial amounts have to be paid for imports, royalties and inflated prices for spare parts, thus further draining away the resources of the Third World. Many Third World countries are sinking into the quagmire of the external debt/foreign technology trap.

The New GATT Agreement: Free Trade vs. Fair Trade

This deteriorating situation is being exacerbated by proposals in the current Uruguay Round of GATT. The United States and other developed countries intend to expand the powers of GATT (which formerly dealt only with the regulation of trade in goods) to include service industries. The major areas proposed for inclusion are banking, insurance, information and commu-nications, the media, professional services such as lawyers, doctors, tourist agencies, accountants and advertising.

Manufacturing and agriculture in many Third World coun-tries are already largely controlled by transnational companies, either through investments or through purchases of their prod-ucts from the world market. It is the service sector in the Third World which still remains basically in the hands of local compa-nies. But as a result of GATT, we can predict that many of the service industries in the Third World will come under the direct control of transnational service corporations within a few years. This would mean the eradication of the last sectors in the Third World which are still controlled by national companies.

These multinational service companies will not only be given the freedom to trade and invest in the Third World, but will benefit from an additional clause called 'national treatment'. This means that any foreign company which wants to set up a base in the Third World in services should be given the freedom to do so and should be treated on terms which are no less favourable than those accorded a national or local company. For example, some Third World countries restrict the participation of foreign banks in the economy by giving only a limited number of licences to such banks, by allowing foreign banks to participate only in certain kinds of banking, or by prohibiting them from setting up branches in small towns so that local banks will have more of the deposit business. Now (under GATT), foreign banks are to be given total freedom; they will be treated just like a local company. We are going to see the marginalisation of local banks and the marginalisation of local financial and professional services.

If one looks very deeply into these processes, one can see that not only economic sovereignty and autonomy are at stake but the very cultures of the people of the Third World as well. Media companies and media owners in the United States or Australia may be given the freedom to set up businesses or to buy out media companies in the Third World—including television and the print media—thereby giving them a powerful influence over the cultures of Third World countries.

GATT will also affect people's health in the Third World. There is already a very big push by the commercial healthcare industry and the insurance companies of the Northern countries for the commercialisation of healthcare services in the Third World. The insurance companies and the big hospital establishments of the North are beginning to buy up hospitals and accelerate the process of commercialisation of Third World healthcare.

Third World countries may be under the impression that, if they give way to the developed countries in areas like services, investments and intellectual property rights, they may benefit in other areas. For instance, they may be given better access to the markets of the industrialised countries through lower tariffs. This may be only an illusion. The industrialised countries have violated similar bargains with Third World countries in the past.

Under GATT, it is not clear to what extent governments—not only in the Third World but even in the United States—will have the autonomy to establish environmental, occupational health and other safety regulations. Some of these regulations may be considered to be against the principles of free trade and free investment. For instance, a few years ago, Indonesia proposed to ban the export of rattan, which is a very important forest product. It is getting scarcer and Indonesia wanted to retain rattan for domestic use. This of course is welcomed by environmentalists who do not want to see the depletion of forest resources. Immediately, however, the United States and the European Community criticised the Indonesian government and said that the export ban was against the principle of free trade. They accused the Indonesian government of taking protectionist steps, and threatened retaliation.

A government could propose that under GATT rules there should not be international trade in toxic waste or in products which are banned (for sale in countries where they are produced) because they are considered dangerous (like pesticides, drugs and so on). Up until today, however, no developed country has put this on the agenda. Some Third World governments tried to put trade in toxic waste onto the agenda of GATT, but this has been ignored by the developed countries.

GATT could be used to protect the environment, but instead it is being used for the reverse. All responsible citizens in the world should fight against the concept that free trade in all circumstances is necessarily a good thing. We should fight for the principle of fair trade rather than the principle of free trade.

An Alternative Vision

The above analysis clearly show that there must be a radical reshaping of the international economic and financial order so that economic power, wealth and income are more equitably distributed, and so that the developed world will be forced to cut down on its irrationally high consumption levels. If this is done, the level of industrial technology will also be scaled down. There will be no need for the tremendous wastage of energy, raw materials and resources which now go towards the production of superfluous goods simply to keep 'effective demand' pumping and the monstrous economic machine going. If appropriate technology is appropriate for the Third World, it is even more essential as a substitute for environmentally and socially obsolete high-technology in the developed world.

But it is almost impossible to hope that the developed world will do this voluntarily. It will have to be forced to do so, either by a new unity of the Third World in the spirit of OPEC in the 1970s and early 1980s, or by an economic or physical collapse of the system.

In the Third World, there should also be a redistribution of wealth, resources and income, so that farmers have their own land to till and thus do not have to look for employment in timber camps or on transnational company estates. This will enable a redistribution of priorities away from luxury-oriented industries and projects towards the production of basic goods and services. With the poor given more resources, there would be an increasing demand for the production of such basic goods and services. With people given the basic facilities to fend for themselves, at least in terms of food crop production, housing and health facilities, Third World governments can reduce their countries' dependence on the world market to sell their resources.

Thus there can be a progressive reduction in the unecological exploitation of resources. With increasing self-reliance based on income redistribution and the growth of indigenous agriculture and industry, the Third World can also afford to be tough with transnationals; it should insist that those invited adhere to health and safety standards that now prevail in the industrial countries. It can reject the kind of products, technologies, industries and projects which are inappropriate for need-oriented and ecological development.

In development planning, the principle of 'sustainable development' should be adopted: the sparing, optimal use of non-renewable resources, the development of alternative renewable resources, and the creation of technologies, practices and products that are durable and safe and satisfy real needs, so as to minimise waste.

In searching for a new environmental and social order we should also realise that it is in the Third World that the new ecologically sound future of the world can be born. In many parts of the Third World and within each Third World nation there are still large areas of ecologically sound economic and living systems, which have been lost in the developed world. We need to recognise and identify these areas and rediscover the technological and cultural wisdom of our indigenous systems of agriculture, industry, shelter, water and sanitation, medicine and culture.

I do not mean here the unquestioning acceptance of everything traditional in the romantic belief in a past Golden Age. For instance, exploitative feudal or slave-based social systems also made life more difficult in the past. But many indigenous technologies, skills and processes that are harmonious with nature and community are still part-and-parcel of life in the Third World. These indigenous scientific systems are appropriate for sustainable development, and should be accorded their proper recognition and encouraged. They should be upgraded if necessary, but they first have to be saved from being swallowed up by the 'modern' system.

Third World governments and peoples in the developed world

have to reject their obsession with modern technologies, which absorb a bigger and bigger share of surplus and investment funds in projects like giant hydro-dams, nuclear plants and heavy industries which serve luxury needs. We must turn away from the obsession with modern gadgets and products which were created by the rich countries to mop up their excess capacity and their need to meet effective demand.

We need to devise and fight for the adoption of appropriate, ecologically sound and socially equitable policies to fulfill needs such as water, food, health, education and information. We need appropriate technologies and even more so the correct prioritising of what types of consumer products to produce. We can't accept 'appropriate technology' that produces inappropriate products. Products and technologies need to be safe to handle and use, they need to fulfill basic and human needs, and should not degrade or deplete the natural environment. And perhaps the most difficult aspect of the fight is the need to de-brainwash the people in the Third World from the modern culture which has penetrated our societies, so that lifestyles, personal motivations and status structures can be delinked from the system of industrialism, its advertising industry and creation of culture. And so the creation and establishment of a new economic and social order which is based on environmentally sound principles to fulfill human rights and the needs of people is not such an easy task, as we know too well. It may even be an impossible task, a challenge which cynics and even the goodhearted in their quiet moments may feel will end in defeat. Nevertheless, it is the greatest challenge in the world today, for it is tackling the issue of the survival of the human race and of the Earth itself. It is thus a challenge which we in the Third World readily accept. We hope that together with our friends in the developed countries, we will grow in strength to pursue the challenge for as long as required, to build the many paths towards a just and sustainable social and environmental order.

GLOBALISM, BIODIVERSITY AND THE THIRD WORLD

Vandana Shiva

✧

Globalism, in the form of Western solutions to environmental problems, has suddenly emerged as a medicine for ecological disease in the South. The history that is being forgotten is that it was the emergence of an earlier globalism, in the form of colonialism, that initiated this environmental degradation in the first place. When Europe colonised 85% of the world, it began a worldwide rape of the Earth's resources. Previously, indigenous knowledge and social systems had ensured the protection of nature by treating vital natural resources as sacred and as held in common. Western reductionist science emerged as a perfect instrument to remove the barriers of sanctity whilst Western market economies were emerging to transform nature's commons into market commodities.

In the first part of this paper, I take India as an example to show how this indigenous sense of sacred natural resources held in common was broken down by the Western colonial powers. Then I will show how the process is being continued by the independent, Westernised governments which followed, and by major international agencies such as the FAO and the World Bank. In the second part of the paper I analyse in some detail the attrition and exploitation of the Earth's biodiversity—'the last frontier' of sacred resources held in common—by powerful commercial interests in the First World seeking to exploit the growing and lucrative biotechnology market.

Science versus nature

Bacon, writing in the late 16th century, has been called the 'father' of modern science. In Bacon's experimental method, there was a dichotomising between male and female, mind and matter, objective and subjective, rational and emotional. Scientific knowledge and its consequent mechanical inventions do not 'merely exert a gentle guidance over nature's course; they have the power to conquer and subdue her, to shake her to her foundations'.

For Bacon, nature was no longer Mother Nature, but a female nature conquered by an aggressive masculine mind. This was eminently suited to the exploitation imperative for the growth of capitalism, concurrent with the widening horizons of the Westerners.

In contrast to the knowledge system created through the scientific revolution, ecological ways of knowing nature are necessarily participatory. Knowledge is ecological and plural, reflecting both the diversity of natural ecosystems and the diversity in cultures that nature-based living gives rise to. Throughout the world the colonisation of diverse peoples had at its root a forced subjugation of the ecological concepts of nature and of the Earth as the repository of the powers of creation. The symbolism of the Great Mother, creative and protective, has been shared across space and time, and ecology movements in the West today are inspired in large part by the recovery of the concept of Gaia, the Earth Goddess.

Parallel to the destruction of the concept of nature as sacred was the process of the destruction of nature as held in common by and for all. The commons were privatised and people's sustenance bases were appropriated to feed the engine of industrial progress and capital accumulation. The commons, which the Crown in England called 'wastelands', were not really waste. They were productive lands, providing extensive common resources for established peasant communities. The enclosure movement was the watershed which transformed people's relationship to nature and each other. The customary rights of people to use the commons were replaced by laws of private property. The Latin root of 'private' means, in fact, to deprive.

The fate of the forests was similar to that of the pastures. The Crown possessed the forests, while the peasants had common rights to forest produce. With the resource demand for capitalist growth, a policy of deforestation was adopted. The peasants lost their common rights, and the Crown and lords of the manors enclosed their deforested land and parcelled it into large farms for lease. The policy of deforestation and the enclosure of the forest led to the largest single outbreak of popular discontent in the 17th Century. Between 1628 and 1631, entire regions were in a state of rebellion.

Colonial exploitation

This process was replicated a hundred or so years later in the rest of the colonised world. In India, the first Indian Forest Act was passed in 1865 by the Supreme Legislative Council, which authorised the government to declare forests and 'benap', or unmeasured lands, as reserved forests. The introduction of this legislation marks the beginning of what is called the scientific management of forests; it amounted basically to the formalisation of the erosion both of the forests and of the rights of local people to forest produce.

Forest resources—like other vital natural resources—had until then been managed as common resources with strict, though informal, social mechanisms for controlling utilisation to ensure sustained productivity. Besides the large tracts of natural forest that were maintained in this way, village forests and woodlets were also developed through careful selection of appropriate tree species.

Colonial forest management undermined these conservation strategies in two ways. Firstly, changes in the system of land tenure through the introduction of the *zamindari* system transformed village resources into the private property of newly-created landlords. People who satisfied their domestic needs from the collectively-owned village forests and grasslands had now to turn to natural forests. Secondly, large-scale felling of trees in natural forests to satisfy non-local commercial needs—such as shipbuilding for the British navy and sleepers for the expanding railway network—created an extraordinary force for destruction. After about half a century of uncontrolled exploitation, the need for monitoring it was finally realised.

The colonial response consisted of handing over ownership to the state and setting up a forest bureaucracy to regulate commercial exploitation and to conserve forests. What the bureaucracy in practice protected was forest *revenues*, not the forests themselves. This typically colonial interpretation of conservation generated severe conflicts on two levels. On the level of utilisation, the new management system catered only to commercial demands and ignored local basic needs. People were denied their traditional rights, which after prolonged struggles were occasionally granted as favours. On the conservation level, since the new forest management was concerned solely with forest revenues, ecologically unsound silvicultural practices were introduced. This undermined biological productivity of forest areas and transformed renewable resources into non-renewable ones.

Resistance movements

With the reservation of forests and the denial of the people's right of access to them, the villages created resistance movements in all parts of the country. The Indian Forest Act of 1927 sharpened the conflicts, and the 1930s witnessed widespread forest *satyagrahas* as a mode of nonviolent resistance to the new forest laws and policies: villagers ceremonially removed forest produce from the reserved forests to defy laws that denied them the right to these products. These nonviolent protests were suppressed with the might of arms. In Central India, Gond tribals were gunned down for participating in the *satyagraha*. In the Himalayan foothills dozens of unarmed villagers were killed and hundreds injured in Tilari village of Tehri Garhwal on May 30, 1930, when they gathered to protest against the forest laws. After enormous loss of life, the *satyagrahas* were finally successful in regaining some of the traditional rights of the village communities to various forest produce. But the forest policy and its revenue-maximising objective remained unchanged.

In independent India the same colonial forest management policy was continued but enforced with greater ruthlessness, which was justified in the name of development, national interest and economic growth. The threat to survival having become more sinister, the response of the people has changed. Sporadic protests have become organised and sustained movements. Chipko is the most spectacular of these.

Chipko is a people's response to a threat to their survival. Beginning as a local grassroots movement, it has spread into the national and the transnational arenas, challenging global paradigms of resource use. While the first global environment meeting of the élite was taking place in Stockholm in 1972, the first grassroots ecology movement was emerging in remote villages in the southern foothills of the Himalayas. The villages were totally unaware of the global event taking place in Stockholm. Their environmental sensibilities were born of their own experience of increasing floods and landslides caused by deforestation; it was rooted in their cultural heritage, which instilled in them a deep respect for the awe-inspiring mountains and their magnificent forests and rivers.

The first Chipko action occurred in March 1974 in Reni village of the Garhwal Himalaya. A group of village women led by one Gaura Devi hugged the trees, challenging the hired loggers who were about to cut down trees for a sporting goods company. Ecological disasters had prepared the ground for this grassroots movement to protect the forests. For example, in July 1970 the Alaknanda valley, in which Reni village is located, had experienced a disastrous flood when the Alaknanda river inundated 1000 square kilometres of land and washed away a large number of bridges and roads. The Chipko movement spread rapidly through the valley. In February 1978 the activists in Henwal valley saved the Advani forests, and in December 1978 the forests of Badyargad were rescued. The news spread beyond Garhwal and in 1983-84, the Chipko strategy was used by activists of the Western Ghats to save the forests of other ecologically vulnerable mountain systems of India.

'Green globalism'
While Third World ecology movements like Chipko focus on people's rights, global prescriptions focus on international markets as a solution to the environmental crisis. The official

Tropical Forest Action Plan (TFAP), promoted by the World Bank and the UN's Food and Agricultural Organisation (FAO), is supposedly designed to halt rainforest destruction; in fact, it is fatally flawed and will actually accelerate forest loss. The plan is in effect a renewed effort at commercial exploitation of tropical forest. As movements like Chipko have shown, it is commercial forestry that was the root of forest destruction in the Third World. Yet, in the name of saving tropical forests, the World Bank and FAO are planning their final destruction.

In 1986 on the basis of the Indian experience I indicated that the plan would result in deforestation, not conservation. More recently the World Rainforest Movement has shown in a report that most projects under the FAO plan are speeding up rates of deforestation by commercial logging and industrial forestry. The plan proposes a 400-600% increase in industrial forestry in the Nepalese Himalayas. What the FAO/World Bank recommends as tropical forestry 'action' is precisely what people's action is resisting in order to save the forests of Amazonia and the Himalaya. The successful projects that this plan cites for 'greening' the Earth have, in practice, been projects for desertification of the Earth and destitution of the people.

Financing goes to increase industrial forestry plantations at the cost of food and forest needs of local populations. Eucalyptus, a pulpwood species, has been the World Bank's favourite monoculture crop. Single species and single commodity production plantations have been the basis of a green revolution in forestry. And like the Green Revolution in agriculture, monoculture plantations do not recover the life of the Earth, they endanger it. They displace diverse trees species which satisfy local needs for fodder, food, fertiliser, fuel and so on. 'Greening' with eucalyptus ends up being a prescription for desertification, as the high water demand and low nutrient return leaves impoverished soils and dry streams, rivers and wells. Everywhere in the world where eucalyptus plantations have been forced on people, peasants and tribals have resisted these forestry projects.

As in the Green Revolution, forestry projects financed by the World Bank have led to the exchange of ecological regeneration principles for those relying on the investment logic of financial institutions. They have also replaced self-help activity with credits, loans and debt, and local genetic diversity with imported uniformity.

Biological diversity

Severe genetic erosion has taken place in recent years as a result of the above mentioned commercialisation of agriculture and forestry, which introduces genetic uniformity as a feature of production. Genetic erosion is an ecological hazard both because it leads to the extinction of life forms which have value in themselves, and because genetic uniformity breeds ecological vulnerability.

The objective for ecological conservation is to preserve species by incorporating diversity into production processes in agriculture and forestry. The imperative for ecological conservation of biodiversity is therefore to reshape commercial forces. However, there is a second approach to biodiversity which attempts to reshape the conservation movement to the logic of commercial forces.

This second kind of concern for the protection of genetic diversity can be called 'commercialised conservation'. The emergence of new biotechnologies has transformed the genetic richness of this planet into strategic raw material for the industrial production of food, pharmaceuticals, fibres, energy, and so on. Commercialised conservation measures the value of conservation in dollars and justifies it on grounds of its present or future profitability. It fails to see biodiversity as having an inherent ecological value in itself. It does not see the need to put an end to the expansion of production processes that wipe out genetic diversity and views biodiversity conservation only in terms of 'set-asides' and 'reserves'.

The recently circulated draft of a plan for the development of a global strategy for the conservation of biodiversity drawn up by the World Resources Institute (WRI) in collaboration with the International Union for the Conservation of Nature, the World Wide Fund for Nature, and the United Nations

Environmental Programme seems to fall in this category of commercialised conservation strategies.

The World Bank is actively pursuing the goal of a global Biological Diversity Action Plan. Towards this aim, a World Bank Task Force on Biodiversity has been set up. The strategy aims to set the stage for an International Biodiversity Decade, whose programme would integrate scientific, social, and political analyses related to living resource conservation and development.

However, the plan lacks both a political perspective as well as an ecological perspective on biodiversity conservation. It fails to address the causes of genetic erosion in the past, as well as threats posed to the maintenance of biodiversity in the future. The WRI plan states that 'Biodiversity is a global resource. All nations benefit from its conservation, yet the current threats are greatest in the developing countries that have severely restricted financial means available to undertake conservation projects.' A number of misconceptions and misrepresentations are evident within this statement: that *biodiversity is a global resource*; that *all nations benefit equally* from the utilisation of biodiversity; that the threats to biodiversity *arise solely within developing countries*; that *genetic erosion takes place because Third World countries have severely restricted financial means*; and that biodiversity conservation *is dependent primarily on money*.

These assumptions misrepresent the issue of biodiversity conservation in its full historical and political context. Biodiversity is not uniformly distributed across the world. It is concentrated in the tropical countries of the Third World, and is therefore primarily a Third World resource. Calling it a 'global resource' opens the door to expropriation of these resources by the North. The WRI approach paper, as well as World Bank task force country papers, avoid analysing the technological and political forces that have caused genetic erosion by leaving out all discussions of plant biodiversity in agriculture. Discussion is limited to biodiversity in forests. The exclusion of agriculture leaves out two critical aspects of biodiversity conservation. It excludes an understanding of how commercial forces have contributed to the destruction of diversity in the past, and it blocks our perception of how the same commercial forces, with new technologies, need 'commercialised conservation' for their biological raw material supplies.

By neglecting to address issues of biodiversity as diversity of all life forms, the proposed action plan fails to see the necessity of incorporating principles of diversity into agricultural and industrial production. It instead proposes 'set-asides' and 'reserves' of wilderness areas as the primary instrument for conservation. However, it is being increasingly recognised that merely setting aside preserves in the remaining relatively undisturbed ecosystems will no longer suffice, both because such areas are too small to be sustainable and because rapid climatic changes that are taking place through processes like the greenhouse effect make these preserves impossible to maintain.

Genetic uniformity in agriculture
Genetic erosion in the Third World has resulted from development policies which have replaced indigenous agricultural and forestry practices based on genetic diversity with practices based on genetic uniformity. International aid has financed the destruction of genetic diversity through the spread of agricultural programmes like the Green Revolution. Corporations and institutions of the North have used the rich genetic diversity of the South as a free resource and a raw material input for their breeding programmes and seed industry. Genetically uniform seeds are then spread in the Third World as purchased 'inputs' for Green Revolution agriculture. This leads to genetic uniformity on the one hand and the creation of debt and financial dependence on the other. There are thus two kinds of 'donors' in biodiversity action. On the one hand there are the gene-rich Third World countries which have donated their genetic resources freely to Northern corporations or to public institutions controlled by the North. On the other hand, financial institutions in the North provide credit to Third World countries for inducing the switch from the use of freely accessible biodiversity to high-priced genetic uniformity purchased from TNCs.

All this is done in the name of agricultural or forestry development.

Agencies like the World Bank—which are now launching the biodiversity action plans—have for the past fifty years been financing the destruction of genetic diversity in the Third World. The World Bank has directly financed the replacement of genetically diverse cropping systems in the Third World with the ecologically vulnerable, genetically uniform monocultures of the Green Revolution. It has also contributed to genetic erosion through the centralised research institution controlled by CGIAR (the Consultative Group of International Agricultural Research) which was launched in 1970 on World Bank initiative and has its offices in the World Bank.

Centralised research and genetic uniformity go hand in hand in agriculture. It is therefore no surprise that the Centre for Wheat and Maize Improvement in Mexico became an instrument for the destruction of genetic diversity of maize, and that the International Rice Research Institute in the Philippines became an instrument for the destruction of genetic diversity of rice in Asia.

The growth of such international institutes was based on the erosion of the decentralised knowledge systems of Third World peasants and Third World research institutes. The centralised control of knowledge and genetic resources was not achieved without resistance. In Mexico, peasant unions protested against it. Students and professors at Mexico's national Agricultural College in Chapingo went on strike to demand a programme different from the one that emerged from the American strategy and more suitable to the small-scale poor farmers and to the diversity of Mexican agriculture.

These institutions and programmes—which the World Bank has launched for conservation—have contributed to genetic erosion. In fact, the continued spread of genetic uniformity is perversely viewed as a means for 'biodiversity' in programmes of the World Bank. As an illustration, John Spears of the World Bank has recommended intensification of monoculture practices in agriculture and forestry for 'preserving biological diversity' in

Asia. This schizophrenic approach to biodiversity—which adopts a policy of destruction of diversity in production processes and a policy of preservation in 'set-asides'—cannot be effective in conservation of species diversity. Biodiversity cannot be conserved unless production itself is based on a policy of preserving diversity.

Thus, World Bank financed 'social forestry' projects in India are, in effect, the replacement of genetically diverse agro-forestry species and crops with monocultures of eucalyptus plantations for the paper and pulp industry. As mentioned above, the globally launched Task Force on Biodiversity is similarly leading to the loss of genetic diversity worldwide.

The genetic resource drain

The International Bureau for Plant Genetic Resources (IBPGR), run by the CGIAR system, was specifically created for the collection and conservation of genetic resources. However, it has emerged as an instrument for the transfer of resources from the South to the North. While most genetic diversity lies in the South, of the 127 base collections of IBPGR, 81 are in the industrialised countries, 29 are in the CGIAR system (which is controlled by the governments and corporations of the industrialised countries) and only 17 are in national collections of developing countries. As Jack Kloppenberg has observed, 'There is empirical justification for the characterisation of the North as a finance-rich but "gene-poor" recipient of genetic largesse from the poor but "gene-rich" South.'

The biodiversity action plan could turn out to be a larger version of the South-to-North transfer of genetic resources that has taken place through institutions like IBPGR. The WRI, in fact, cites the IBPGR experience as a model for biodiversity action in its Tropical Conservation Financing Project. It states that the IBPGR works to 'ensure the collection and conservation and use of germ plasm so as to contribute to raising the standard of living and welfare throughout the world.' Experience, however, shows that IBPGR has not contributed to equal benefits worldwide, but has used Third World resources for the benefit of industrialised countries.

Biotechnology

With the emergence of the new biotechnologies, the polarisation between the North and South around issues of biodiversity will probably be aggravated, since the North will try to continue to treat the biodiversity of the South as a freely accessible global resource while it attempts to privatise genetic resources through patent laws and intellectual property protection related to life forms. No biodiversity action plan can hope to reverse the threat of species extinction without a serious assessment of the causes of genetic erosion in the past, and the political threats to biodiversity in the future.

Biotechnology is being viewed by many environmentalists as a solution to the ecological problem of genetic erosion. Gus Speth of the World Resources Institute states that the 'world's emerging biotech industry provides many of the tools needed for environmentally sustainable growth.' However, the commercialisation of biodiversity through the biotech industry and its commercialised conservation creates new threats of ecological vulnerability, and new threats of political polarisation between the North and the South.

Corporate interests view patent protection as a prerequisite for innovations in biotechnologies. However, in the area of life forms, the granting of industrial patents is full of risks and controversies.

The first controversy in this area is related to ethics and respect for the integrity of life. As animal rights activist Joyce D'Silva has stated, 'What happened to our ethics? When I read statements from biotechnology researchers such as "transgenic animals can also be viewed as production systems for useful pesticides" or "new animals ought to be patentable for the same reason that robots ought to be patentable because they are both products of human ingenuity," I am worried. I believe that is unethical, profoundly so. Animals should not be viewed only from the point of view of human usefulness and profit.'

The main problem with viewing biotech as a miracle solution

to the biodiversity crisis is related to the fact that biotechnologies are, in essence, technologies for the breeding of uniformity in plants and animals. Biotech corporations do in fact talk of contributing to genetic diversity. As John Duesing of Ciba Geigy states, 'Patent protection will serve to stimulate the development of competing and diverse genetic solutions, with access to these diverse solutions ensured by free market forces at work in biotech ecology and seed industries.' However, the 'diversity' of corporate strategies and the diversity of life forms on this planet are not the same thing, and corporate competition can hardly be treated as a substitute for nature's evolution in the creation of genetic diversity.

In fact, it is the commercial logic of profit maximisation that is the primary cause for species extinction. The genetically engineered products of corporate biotechnology ventures will not only be genetically uniform and ecologically fragile in themselves, they will pose new ecological threats to other life forms. Introduced species usually have ecologically disruptive effects. The release of genetically modified plants and microorganisms into the environment threaten to become a new kind of environmental bio-hazard.

Profiting from genes

The issue of patent protection for modified life forms raises a number of unresolved political questions about ownership and control of genetic resources. The problem is that in manipulating life forms you do not start from nothing, but from pre-existing life forms which belong to others—perhaps through customary law. Genetic engineering does not create new genes, it merely relocates genes already existing in organisms. In making genes the object of value through the patent system, a dangerous shift takes place in the approach to genetic resources.

Most Third World countries view genetic resources as a common heritage. In most countries animals and plants were excluded from the patent system until recently, when the advent of biotechnologies changed concepts of ownership of life. With the new biotechnologies, life itself can now be owned. The potential for gene manipulation reduces the organism to its genetic constituents. Centuries of innovation are totally devalued to give monopoly rights over life forms to those who manipulate genes with new technologies. The intellectual contribution of Third World farmers in the areas of conservation, breeding, domestication and development of plant and animal genetic resources—over many generations and for thousands of years— is entirely disregarded. The argument that intellectual property is only recognisable when performed in laboratories with white lab coats is fundamentally a racist view of scientific development.

Patenting genes makes biology stand on its head. Complex organisms—which have evolved over millennia in nature and through the contributions of Third World peasants, tribals and healers—are reduced to their parts, and treated as mere inputs into genetic engineering. Patenting of genes thus leads to a devaluation of life forms by reducing them to their constituents and allowing them to be repeatedly owned as private property. This reductionism and fragmentation might be convenient for commercial concerns, but it violates the integrity of life as well as the common property rights of Third World peoples. On these false notions of genetic resources and their ownership through intellectual property rights are based the 'bio-battles' at FAO and the trade wars at GATT.

The US government is using trade as a means of enforcing its patent laws and disregarding intellectual property rights of the sovereign nations of the Third World. The US government has accused countries of the Third World of engaging in 'unfair trading practice' if they fail to adopt US patent laws which allow monopoly rights in life forms. Yet it is the US government which has engaged in unfair practices related to the use of Third World genetic resources. It has freely taken the biological diversity of the Third World to spin millions of dollars of profits, none of which have been shared with Third world countries, the original owners of the germ plasm. For example, a wild tomato variety (*Lycopresicon chomrelewskii*) taken from Peru in 1962 has contributed $8 million a year to the American tomato processing industry by increasing the content of soluble solids. Yet none of these profits or benefits have been shared with Peru, the original source of the genetic material.

According to Prescott-Allen, wild varieties contributed $340 million per year between 1976 and 1980 to the US farm economy. The total contribution of wild germ plasm to the American economy has been $66 billion, which is more than the total international debt of Mexico and the Philippines combined. This wild material is 'owned' by sovereign states and by local people.

As drug companies realise that nature holds rich sources of profit they have begun to covet the potential wealth of tropical moist forest as a source for medicines. For instance, the periwinkle plant from Madagascar is the source of at least 60 alkaloids which can treat childhood leukemia and Hodgkin's Disease. Drugs derived from this plant bring in about $160 million worth of sales each year. Another plant, from India (*Rauwolfa serpentina*), is the base for drugs which sell up to $260 million a year in the United States alone.

Unfortunately, it has been estimated that with the present rate of destruction of tropical forests, 20-25% of the world's plant species will be lost by the year 2000. Consequently, major pharmaceutical companies are now screening and collecting natural plants through contracted third parties. For instance, a British company, Biotics, is a commercial broker known for supplying exotic plants for pharmaceutical screening by inadequately compensating the Third World countries of origin. The company's officials have actually admitted that many drug companies prefer 'sneaking plants' out of the Third World rather than going through legitimate negotiating channels.

The United States National Cancer Institute has sponsored the single largest tropical plant collecting effort by recruiting the assistance of ethnobotanists who in turn siphon off the traditional knowledge of indigenous peoples without any compensation. Estimates of the value of the South's germ plasm for the pharmaceutical industry ranges from an estimated $1.7 billion now to $17 billion by the year 2000.

Third World biodiversity has made an immeasurable contribution to the wealth of the industrialised countries, yet corporations, governments and agencies of the North continue to create legal and political frameworks to make the Third World pay for what it originally gifted. The emerging trends in global trade and technology work inherently against justice and ecological sustainability. They threaten to create a new era of bio-imperialism, built on the biological impoverishment of the Third World and the biosphere.

The intensity of this assault against Third World genetic resources can be seen from the pressure exerted by major drug and agricultural companies and their home governments on international institutions such as GATT and the FAO to recognise such resources as a 'universal heritage' in order to guarantee them free access to these raw materials. International patents and licensing agreements will increasingly be used to secure a monopoly over valuable genetic materials which can be developed into drugs, food and energy sources.

The Third World must urgently take stock of its genetic resources, particularly those contained in tropical forests. Rather than permit the North to 'rescue' the world's tropical forests for their own economic interests, conservation measures must be undertaken for the long-term benefit of the Third World and indigenous peoples. Due respect and recognition must be accorded to the knowledge and interests of indigenous peoples.

A 'People's Biodiversity Conservation Action Plan'
In contrast to the commercialised conservation plans for biodiversity, a 'people's biodiversity conservation plan' needs to take account of the following:

- Regenerating diversity must be the basis not only of conservation, but also of production in agriculture, forestry, energy and health care.
- The practice of diversity can only be ensured through decentralisation. Centralised systems of research, production and conservation force the spread of genetic uniformity and genetic erosion.

- The practice of diversity has been characteristic of indigenous systems of production in the Third World. Biodiversity conservation plans need to contribute to the regeneration of these systems.
- The knowledge and intellectual contributions of generations of Third World 'innovators', peasants, tribals and traditional healers need to be recognised and treated on an equal footing with innovation in the labs of industrialised countries to correct the distortions and inequities being introduced through patenting of life forms.
- The contribution of Third World germ plasm from wild as well as cultivated varieties to capital accumulation in industrialised countries needs to be recognised and compensated for in a just and ecological manner, not merely as tokenism. There is injustice inherent in current technological and trade practices, which treat genetic resources which come from the Third World as freely available while the same genetic material used by scientists and corporations in the North is protected by patents, treated as private property and sold back to the Third World at exorbitant costs.
- We must address the issue of 'ownership' of life through patents on life forms, with all its ethical, legal and political implications.
- Third World countries should prohibit all researchers, social scientists and scientists who are working for foreign interests from conducting research on and/or collecting genetic resources. Existing contracts or agreements to do research, screening and collecting of genetic germ plasm should be terminated to stop the transfer of valuable germ plasm to the North and to safeguard the heritage of the Third World.
- All countries should introduce legislation and institutional safeguards to protect genetic resources.
- The activities of all transnational corporations in this field need to be systematically monitored.
- Countries should also systematically monitor and analyse the activities of international agencies like GATT, FAO, the World Intellectual Property Organisation and the International

Union for the Protection of New Varieties of Plants. These international organisations are dominated by the Northern countries and have been used by them to rob the Third World of its resources and the rights of Third World peoples.

- We need to encourage and provide incentives for local research, identification and documentation of genetic resources, and should set up national gene banks free from transnational corporation and foreign government funding, technical assistance, control and involvement. South-to-South cooperation and assistance in the setting up of gene banks and research should be encouraged.

DEVELOPMENT FALLACIES

Edward Goldsmith

✧

The 'development' currently imposed by the industrialised nations on the Third World is producing a whole series of interconnected negative impacts on the very people the process purports to help. In this paper, I will discuss two important strands in this complex weave of causes and effects: how 'aid' mainly benefits the industrialised world via the opening up of Third World markets for its manufactured goods, and the connections between development and population increase in the Third World.

The Fallacy of Aid

Those with a superficial knowledge of the development process often remain convinced that aid is designed to help the peoples of the Third World. Even many environmental institutions still appear to believe this and persist in campaigning for increased aid. Yet, surely, if the governments of the industrial countries were really concerned with the welfare of the people of the Third World, they would have provided some of their vast food surpluses (which cost hundreds of millions of dollars to store) to the starving people of Africa—even if this would not have solved any long-term problems. Alternatively, they could have spent on famine relief the money which US farmers are now paid not to produce food.

Needless to say, no politician has suggested we do anything of the sort. On the contrary, in Britain in 1985-86, in the face of the worst and most widespread famine Africa has ever known, our government actually reduced its aid to the people of that continent, so that, as John Madeley notes, 'There is more in the kitty for better off countries such as Turkey and Mexico' (which, unlike the countries of Africa, have the money with which to buy British manufactured goods).

This ability to buy goods from the industrialised countries is the crux of the matter. Indeed, the US Department of Agriculture admits that American food aid is a means of creating a demand for imports from the US. 'Food aid,' it declares, 'can pave the way for US commercial exports.' For example, in 1956-58, United States food aid to 17 overseas markets was $3.1 billion, and commercial sales of all goods was $3.6 billion. Two decades later, food aid from the United States to these same countries was only $756 million, and commercial sales had grown to $43 billion.

Aid and trade

One of the main reasons why aid is sound commercial practice is that much of it is officially tied to sales of manufactured goods. In the same way that colonies were once forced to buy their manufactured goods from the country that had colonised them, today's recipients of aid must spend much of the money they receive (money that is supposed to relieve poverty and malnutrition) on irrelevant manufactured goods that are produced by the donor countries. What is more, if they dare refuse to buy any of our manufactured goods or to sell us some resource—generally, because they want to keep it for themselves or to conserve it for the future—they are immediately brought to heel by the simple expedient of threatening to cut off further aid, on which they have become increasingly dependent.

Thus, a few years ago, a World Health Organisation study revealed that only a minute fraction of commercial pharmaceutical preparations were of any real therapeutic use. Bangladesh, one of the poorest countries of the world, decided to take the study seriously and announced that it would ban all superfluous drugs. The US government immediately reacted by threatening to withhold food aid if Bangladesh discriminated in this way against US pharmaceutical manufacturers. So too, in 1979, the Bangladesh government decided to stop selling rhesus monkeys (a threatened species) to a US company called Mol Enterprises for experimentation in its laboratories. The US government's response was, as *New Scientist* notes, 'swift and strong' and 'even included a suggestion that American aid could be cut off if Bangladesh refused to honour its contract with Mol Enterprises, the monkey importers.'

The British government behaved in a similar manner with the government of India by threatening to cut off aid if India did not go ahead with plans to buy 21 Westland helicopters at a cost of £60 million—an effort which, it is encouraging to note, was bitterly opposed by responsible elements within the Overseas Development Administration.

All this is simply a slightly more sophisticated means of achieving what Commodore Perry achieved by bombarding Nagasaki in order to force the Japanese to trade with America, and what Britain achieved by going to war with China so as to force it to buy opium from British merchants in India.

Bretton Woods

It was at the Bretton Woods Conference in 1944, held under US leadership, that aid was institutionalised as the industrialised world's principal tool of economic colonialism. At that conference, 44 nations agreed to set up the key international institutions. They were: the International Monetary Fund (IMF); the World Bank (IBRD); and the General Agreement on Tariffs and Trade (GATT). These highly interconnected 'agencies' formed a single integrative structure for manipulating world trade, which until the early seventies was basically dominated by the United States of America. The original role of the IMF was to make sure that member nations pegged their currency to the US dollar or to gold (72 percent of world gold supplies were in the possession of the US). This expedient would, among other things, make it difficult for Third World debtors to get out of their financial obligation to the Western banking system by manipulating their currencies.

The World Bank's first function was to reconstruct Europe's shattered economy after World War II. Its second function was to prevent the recurrence of a 1929-style slump by systematically expanding the Western economy. Significantly, as Susan George writes in *A Fate Worse than Debt*, Article 1 of the IMF charter describes six objectives, the principle being: 'to facilitate balanced growth of international trade and, through this, contribute to high levels of employment and real income and the development of productive capacity... To seek the elimination of exchange restrictions that hinder the growth of world trade.'

She goes on to comment, 'Even those objectives described in the first Article that may appear strictly financial are, in fact, geared to a single overriding objective: the growth and development of world trade.'

As a result, the World Bank soon moved into the business of Third World development (its main activity). It has built roads, harbours, ports, and so on—to supply the infrastructure required to enable the importation of manufactured products and the export of raw materials and agricultural produce. It then invested heavily in energy generation, in particular in hydropower, the adverse consequences of which have been documented in our book *The Social and Environmental Effects of Large Dams*.

More recently, since the 1970s, the Bank has played a leading role in financing the commercialisation of agriculture in the Third World and, in particular, the substitution of export-oriented plantations and livestock-rearing schemes for traditional subsistence farming designed to feed local people. In doing this, it has made a massive contribution to the growth of poverty and famine in Africa and South and Southeast Asia.

The role of GATT, the third of the institutions set up at Bretton Woods, was to liberalise trade and, hence, to ensure that Third World countries did not try to manufacture locally what they could buy from Western countries—that is, indulge in highly frowned-upon 'import substitution'.

IMF conditionalities
The IMF has complemented the work of GATT in this respect. Loans, either from the IMF itself or the World Bank, have only been provided to governments that have undertaken to observe

the IMF 'conditionalities'. This has meant above all, abolishing import quotas and reducing import tariffs to a minimum. This prevents Third World countries from protecting their fledgling industries against competition from the established and highly capitalised enterprises of the industrial world—industries that during the early stages of their own development were themselves well-protected from foreign competition, and many of which still are.

Third World governments have also been required to devalue their currencies to make their exports more attractive to the industrialised countries—which means they must pay more for their imports. They are also required to abolish expenditure on social welfare and, in particular, on food subsidies which are often badly needed to protect the mass of the population from the disruptive effects of the rapid socio-economic changes that development inevitably brings about. Such expenditure is seen as being better spent on Western imports or on building up a country's industrial infrastructure.

If the Fund were really interested in the fate of the people of the Third World, it would not cut down on food subsidies to the poorest people of the world, most of whom only need food handouts because they have been deprived of their land to make way for large-scale development schemes (largely funded by the West), and the import of nonessential items—armaments being a prime example. Yet, as Susan George notes, 'The IMF consistently demands that its pupils make drastic reductions in civil spending, but arms budgets remain untouched. When asked about this anomaly, Fund personnel recoil and explain in pained tones that such measures would be "interfering in the internal affairs of sovereign nations" (which is exactly what the Fund does every working day).'

Similarly, the IMF could insist on a purge on corruption in Third World governments and, in particular, 'capital flight', which could be responsible for the loss of as much as $100 billion a year.

Apart from being made to devalue their currencies and cut social expenditure, Third World governments must also undertake to mechanise their food production—that is, to adopt the

'Green Revolution', thus providing an important market for Western agricultural machinery and agro-chemicals. They must also replace subsistence agriculture with export-oriented agriculture so as to provide the West with agricultural produce. (Third World countries must export in any case to pay for the capital equipment they need to mechanise their agriculture and to finance the mass of manufactured goods that now flows into their countries.)

This package of policy prescriptions has been imposed on Third World countries by all the multilateral development banks. Rupesinghe, for example, quotes a report by the Asian Development Bank on SE Asia's economy: 'Countries must move away from inefficient import substitution policies and free the economy of import controls and price controls. The Green Revolution must be promoted as a "genuine dynamic force" of economic development. Agribusiness should be invited to cooperate in a country's drive towards self-sufficiency. Resource allocations must shift from domestic production to export crops for the world market. Local support, generous tax incentives, profit registrations, should be provided for foreign investors, and legislation must be enacted to create a climate of stability for foreign investment.'

Recycling capital

Since the early seventies, the amount of capital pumped into the Third World to finance such development policies has increased massively, as has the destruction it has financed. One reason for this capital expenditure is the need to recycle the vast sums of money accumulated by the OPEC countries back into the Western economic system.

This is fully admitted by the US government in one of its publications: 'In the 1970s the large increase in petroleum prices gave rise to large amounts of what were called petrodollars, since petroleum was (and still is) paid for in dollars. Commercial bankers were enjoined by the United States and international agencies such as the International Monetary Fund to reloan or recycle these dollars to keep the international economy from collapsing. This they did to a fault, giving rise to what later came to be known as the international debt crisis.'

Unfortunately, the process is about to be repeated, since, with the aid of the World Bank, we now plan to recycle, via the economies of Third World countries, Japan's annual $80 billion surplus—which is equivalent to the OPEC surpluses of the late 1970s.

The impact that the vast development schemes (which alone can sop up all that money) must inevitably have on the already devastated environment of the Third World is too awful to contemplate.

The Fallacy of Development and Population Stability

An essential reason why economic development cannot help combat malnutrition and famine is that it must inevitably give rise to a population explosion. The experience has been the same everywhere. As soon as a traditional society embarks on the path of economic development, its population simply explodes. It happened in Britain, where the population was under 8 million when the Industrial Revolution began, and where it increased by more than 7 times before it eventually stabilised. It is happening today wherever economic development occurs throughout the Third World.

Our reaction to this problem is always the same. Population growth is interpreted in such a way as to make it appear amenable to a technological solution—the only solution the North is organised to provide. It is the only solution that involves producing the sort of hardware that the corporations into which our society is organised can manufacture; the only type of solution, in fact, that is 'economic' and hence politically acceptable.

The World Bank estimates that to achieve 'a rapid fertility decline in Sub-Saharan Africa would mean increasing the amount of money spent on 'family planning' by 20 times by the end of the century.' Just think how the export of contraceptive pills, condoms, IUDs and other forms of birth-control gadgetry will rocket. Is it possible to imagine a more 'economically viable' and 'politically expedient' solution? But what is the point of providing vast numbers of men and women with expensive birth-control devices if, as happens to be the case, they want the children whose birth these devices are designed to prevent? The answer is clearly none at all.

Stable populations

We tend to forget that the populations of traditional societies were stable for centuries, if not millennia. They had to be, in order to preserve their social structure and their physical environment. The reasons for that stability are clear.

To begin with, traditional societies exploited a wide range of cultural strategies—such as taboos against sexual activity during lactation and during the first years of widowhood, or the prohibition against widow remarriage among certain castes in India—which are intended to minimise population growth. However, as a society's social structure and cultural pattern are destroyed by the process of economic development, such population control strategies can no longer operate, which means that the population in question simply grows out of control.

The population of traditional societies is stable for another reason, namely that each individual belongs to an extended family and lineage group which provide an extraordinary degree of security. What is more, each individual has a right to the land they and their family occupy by virtue of their status as a member of these groupings. In addition, the agricultural methods used are designed to maximise security even at the cost of limiting yields.

Development changes all that. In fact, it shatters every aspect of traditional life. Indeed, it is a process which, as Robertson notes, is 'more likely to generate unhappiness, violence and tyranny than social harmony.' Esenstadt also considers that because 'modernisation entails continual changes in all major spheres of a society, [this] means, of necessity, that it involves processes of disorganisation and dislocation, with the continual development of social problems, cleavages and conflicts between various groups and movements of protest and resistance to change. Disorganisation and dislocation thus constitute a basic part of modernisation, and every modern and modernising society has to cope with them.'

In particular, development destroys a society's cultural pattern and its associated social structure. The society thus disintegrates and becomes atomised, as in the industrialised world today. Such a society can no longer govern itself, nor provide its members with the security that it previously provided: instead it must now be governed by a government bureaucracy, which previously would have no *raison d'etre*. Such a bureaucracy, however, can never compensate people for the inestimable social capital provided to them by the social groupings to which they previously belonged. Nor can participation in the formal economy, usually as grossly underpaid casual workers, compensate people for the loss of their land—which is inevitably taken over to accommodate more economic land uses. All this creates the most terrible misery and insecurity, and in order to survive, people are forced to seek an alternative strategy for providing themselves with some sort of security, however precarious. One such strategy is to have more children, who can be hired out as labourers or who can even be trained to beg and steal in the cities.

Malthusian dogma and the 'demographic transition'

Interestingly enough, one of the official explanations of the population explosion is that, with development, food production increases, which means that more food is available to the local population which, in a true Malthusian manner, can be counted upon to breed up to the available food supplies. The opposite, however, seems to be true. Thus, although food production has increased in, for example, both India and Zimbabwe in the last decade, this has not meant that more food has been made available to the local population. On the contrary, the food has mainly been produced for export or for consumption in the cities and, in reality, as Banerjee and Kothari and Jackson point out, less food is available to the rural masses. This was also the case in Ireland, when, during the eighteenth and early nineteenth centuries, the population exploded from two to eight million. During that period much of the arable land was taken away from the peasants by the big estates, with the result that the peasants had to rely on the potato, the only crop that could feed a family from the small area of degraded land that remained at its disposal. During the course of the 19th Century, an increasing proportion of food produced in Ireland was exported to England; the exports were in no way reduced during the famine, which killed something like two million people and forced another two million to seek refuge beyond the seas.

The truth seems to be that, in an atomised society at least, the population explodes not when there is more food to eat (as conventional wisdom tells us) but when, on the contrary, there is less food to eat.

Of course, we are assured that development will provide people with a new form of security, one provided by membership in the growing formal economy. As people become more secure, we are told, they will then have fewer children, as has happened in the West. What the development industry does not tell us, however, is that it is economic development that created the insecurity in the first place.

To assume that the 'demographic transition' will occur in the Third World is in any case an act of faith. We are not at all sure why the population rate has fallen in the West. Is the fall in fact due to increased security? Or are other factors implicated, such as the fear of the future which looks ever grimmer? Or even the serious pollution of human spermatozoa which has radically reduced the sperm count of young males in the Western world and made a considerable proportion of them 'functionally sterile'?

Moreover, the economy of the Third World is not expanding nearly fast enough to absorb the growing hordes of unemployed in the cities and is never likely to; hence the security that participation in the formal economy could provide is available to an ever smaller proportion of the population. Indeed, the Third World can never conceivably attain the level of material prosperity we know at present in the West, which has indeed been associated (however superficially) with a reduction in fertility.

What is certain is that the much anticipated 'demographic transition' is not occurring in the Third World today. As Lester Brown notes: 'The "demographic transition" that has marked the advance of all developed countries may be reversed for the first time in modern history. African countries have now moved beyond the first stage of this transition, with the equilibrium between high birth and death rates. But virtually all remain stuck in the second stage, with high birth rates and low death rates. In this stage, population growth typically peaks at three percent or so per year.'

If the rate of population growth has fallen slightly in the dry tropics today, it is probably because of an increased death rate from famine, malnutrition and the diseases to which underfed people are particularly vulnerable. Indeed, it is only through such crude means that development can help control the very population explosion that it has itself brought about.

Even if the demographic transition did occur, it could not conceivably solve the real problem. A growing population is not intolerable *per se* but because of the increasing impact it must have on the natural environment. This impact is greatly magnified by the increase in material consumption made possible by economic development. To seek to reduce population by systematically encouraging economic development is thus self-defeating since it can only increase consumption and thus environmental destructiveness.

THE PHILIPPINE
ENVIRONMENTAL CRISIS

Filipina del Rosario-Santos

✧

The Philippines' environment has been suffering from a serious crisis that suddenly everyone seems concerned about. Such concern is understandable, given the extent of the problem. Its proportions have reached the point where not only is the future's endowment at stake, but even the present populace is suffering from catastrophes brought about by the rape of our ecosystem.

This paper will examine the most serious environmental problems now facing the Philippines. It is worthwhile to analyse them closely, in order to come up with concrete programmes that can resolve this crisis in a holistic manner.

Loss of resources
Of the 10 million hectares of dipterocarp rainforests the Philippines had in 1950, less than one million remain. Our total forest cover was 16 million hectares in 1968 (54% of total land area), and is today down to less than 6.5 million hectares (22% of total land area). Of the Philippines' 73 provinces, only four still have at least 40% forest cover.

At present, the total volume of trees exceeding 15 cm. diameter (breast height) is about 750 million cubic metres (cu.m.), about half being hardwoods. The volume exceeding the harvestable size of 55 cm. is, however, only 260 million cu.m., of which only 180 million cu.m. are hardwoods.

At present, the annual harvest is about 3.5 to 14.0 million cu.m. per year; adding the amount of illegal or unlicensed logging might well double this figure. If present harvest rates continue, the present forest stands will not last more than 35 years. Hardwood forests planted today would not be mature enough for a first cut before present resources are exhausted.

The rate of deforestation during the 1980-88 period is estimated at 119,450 hectares per year, although some estimates place it as high as 400,000 hectares per year. This large-scale deforestation has led to several subsidiary problems.

About 1 billion cubic metres of topsoil—essential for agriculture—are lost to erosion every year. This runoff (equivalent to 100,000 hectares of land to a depth of one metre) eventually results in incalculable losses in standing crops, silting of rivers and lakes, damage to coral reefs, floods and drought, clogging of irrigation channels, damage to infrastructure works and a reduction in the quality of water for industrial and home use.

Thirteen provinces have already suffered severe erosion, with 50 to 83 percent of their total land area eroded. Thirteen other provinces have undergone erosion nearly as severe, with 40 to 50 percent of total area already eroded. Pantabangan dam has lost 60 years of service life due to heavy siltation, while Magat dam has lost 40 years. Unproductive grasslands of *cogon* (imperata cylindrica) and *bag okbok* (themeda triandra) occur in about 5 million hectares, often in critical watersheds.

Forests influence climate and weather. They also absorb rainwater before it courses down mountain slopes along rivers and streams that fill lakes or empty into the sea. Catastrophic floods have been occurring even in areas where these phenomena were—until recently—almost unheard of. Droughts, which were unknown in such places as Mindanao, are now a periodic phenomenon. Lack of rain has critically lowered the water level of Pantabangan dam in recent times, while in the rainy season rivers overflow their banks.

Biological diversity

Our country's 7,100 islands are home to an exceptionally rich assemblage of both terrestrial and marine life forms. There are estimated to be about 12,000 plant species in the Philippines, of which 3,800 are found nowhere else in the world. There are at least 170,000 species of fauna, mostly insects and mostly unidentified. Terrestrial vertebrate species number over 800, of which

600 are birds and 210 are mammals.

A number of our wildlife plant and animal species are facing extinction. Out of the 1,657 known wildlife species, at least 46 species have already been rendered rare and endangered. The list includes the Philippine eagle, tamaraw, tarsier, dugong, Philippine crocodile and the monitor lizard. Forest plant species on the endangered species list include seven ornamental plants. These species contain genetic potentials that may supply some of man's needs in the future. They also represent a biological heritage which future generations may only be able see in pictures.

Destruction of marginal lands

About 5 million inhabitants of diverse traditional communities are in the uplands—representing almost one third of the present upland population. Most of them are dependent on the forest for their livelihood. Their production system and their mode of resource ownership and access are critically woven into their political, social and religious customs. Baring of the uplands for timber has forced changes in the lifestyle and culture of these communities. It has meant economic dislocation. But more importantly, timber extraction has condemned these people to spiritual impoverishment.

Marginal lands are converted forests, grasslands, brushlands and barren areas. These areas have infertile soil and are more prone to soil erosion because they have relatively steep slopes—18% or more. In recent years, migrants have been slowly invading these steep slopes, evidence that strong pressures are forcing people to leave the lowlands. Farmers in these areas are among the poorest in our population, and many are near the famine level due to frequent crop failures. Specific cultivation techniques and perennial crops must be part of the farming scheme to make these areas productive. Such schemes need security of tenure for the land to be adapted by the farmers and will take a number of years to succeed.

Already, 11.9 million hectares of uplands (about 41% of our total land area) are reportedly under cultivation. Government efforts to reforest these marginal lands have been too slow to match the rate of degradation. Statistics record newly replanted areas of about 60-80,000 hectares during 1978-83. In 1987, only 40,000 hectares were reported reforested.

The upland population has reached about 18 million, or 30% of the national population. About 7.5 million are living within the public forest reserve, and are probably practising farming on steeply sloped areas. With the country's high population growth rates, and a trend towards increasing upland migration, the present upland population is expected to reach about 26 million by the year 2000. Only a programme of genuine land reform and local development can strategically address the growing crisis in the uplands.

Degradation of fishery resources
Consider some of the most notorious cases:

- Lake Laguna is so heavily polluted that the rate of fish production has been drastically reduced, severely affecting the livelihood of lake fishermen. Of the fishes remaining in the lake, most are inedible due to a particular disease called *kurikong ng isda*. Balut production has gone down due to a decrease in the population of ducks and/or a slowed growth rate. The pollutants responsible are fertilisers and pesticides from surrounding agricultural lands, toxic wastes like heavy metals from factories, domestic waste and garbage, and sewage from homes. In addition to pollution, heavy siltation due to runoff from ill-managed watersheds is another serious problem.
- A total of 40 rivers, including all those in Metro Manila are biologically dead due to pollution. A total of 480,800 hectares of fresh water areas are also affected by saltwater intrusion.
- Most of the inland fishery resources have been abused. When fishpen technology was introduced, a good number of our lakes became overcrowded with fishpens. Introduced predator species like *tilapia* and shrimps cause displacement of naturally occurring fishes like *sinar apan* in Lake Buhi and *maliputo* in Lake Taal. In some cases, these fishpens are owned by big

businessmen or the military and therefore have become instruments of oppression of small fishermen.

Saltwater resources are also threatened. The Philippines has a coastline twice as long as that of the United States, and we have one of the richest coral resources in the world. The importance of corals to fisheries are numerous, and include serving as rearing and feeding grounds for diverse fish species; a protective barrier against waves; and a source of pearls and building materials.

With the introduction of destructive fishing techniques such as muroami, cyanide fishing and dynamite blasting, the corals are being destroyed beyond their capacity for regeneration. Siltation due to forest denudation is also a major source of coral degradation. In addition, coral mining and harvesting for use as construction material and for the ornamental shell trade also adds to the stresses on Philippine corals. Despite the 1976 ban on coral trade, it will probably take 40 years or more for dynamited reefs to return to fair condition (defined as at least 50 percent live coral cover.) A survey has revealed that half of the coral stations surveyed have poor reef cover (0-25% live coral.)

Coastal mangrove forests serve as a spawning ground for a great number of fishes. These forests have been cut down at a very fast rate. Approximately 450,000 hectares of mangrove were thought to exist in 1918. By 1980, this had dwindled to 149,000 hectares, or 29% of the 1918 area. The steady reduction in the past several decades can be attributed to the combined impact of forest clearing for fishpond establishment, harvesting mangroves for charcoal or fuel wood production, and—to a lesser extent—harvesting for domestic use. Fishpond conversions are often the final step in a process of incremental destruction, as authorities become convinced that particular mangrove stands are no longer of value. Overharvesting and lack of replanting are usually the initial sources of degradation. The problem is compounded by the lack of government policy enforcers in these areas.

In addition to these factors, overfishing by foreign and large-scale fishing trawlers have placed our coastal fishery resources in a state where irreparable damage will occur if the situation is not properly managed. This is very alarming because fish is one of the cheapest sources of protein for our rapidly growing population.

Finally, the red tide phenomena affecting shell fishes in various coastal areas of Luzon and Visayas are symptoms of pollution of these marine coastal fisheries.

Pollution

Of the 6,521 industrial firms surveyed by the National Pollution Control Commission in 1984, a total of 1,057 (16.2%) cause water pollution. Only 734 (11.3%) had waste water treatment plants.

This problem is compounded by the increasing population density in urban areas, especially Manila. More than 20 million people are now living in urban areas, 8 million of these are in Metro Manila. By the year 2000, the total is expected to reach 30 million, with about 13 million in Manila. About 35% of the Manila population lives below the 1983 absolute poverty threshold.

Only 10-12% of Manila households are served by a wastewater disposal system; most waste water simply runs off into *esteros* and other waterways. Metro Manila generated about 7,000 tons of solid waste daily in 1988 of which only 65% was collected, leaving 2,000 tons daily to be burned, thrown in waterways, or to be mound on the ground.

Modern technology has condemned the Filipino farmer to a perpetual dependence on chemical fertilisers and pesticides. The Masagana 99 and the Maisagana programmes heightened this dependence. High yields require around four bags of fertiliser per hectare. Doubling the yield means a tenfold increase in fertiliser input.

The widespread use of pesticides by farmers in small land holdings in Central Luzon has increased mortality rates by 27%. The vegetable growers of Benguet have been poisoned by pesticides, with the incidence of illness correlating with duration of use. Human milk in some towns of Laguna already contains DDT.

Off-site, agricultural pollution has caused increased eutrophication of bodies of water, resulting in noxious algal blooms that recur in Laguna de Bay. Cases of water pollution are numerous but poorly documented due to the lack of financial support to boost research to the required level.

Population growth

Though the major cause of environmental degradation is still the unequal distribution of wealth, present population growth rates cause additional stresses on the environment. With a growth rate of 2.7% and a doubling time of 17 years, the present population is expected to reach 120 million by the year 2006. An additional 3 million job-seekers will enter the labour force. This is expected to increase the unemployment rate, especially because no serious efforts to industrialise have been undertaken by the past or present regimes.

Given this condition, Filipinos, especially the poor, will be competing for very scarce resources. Already, many are being driven to the uplands due to poverty.

The crisis can only be resolved if a drastic redistribution of resources is undertaken and the poor—more than the present ruling élite—are given more right to the utilisation of natural resources. In addition, a well-planned and well-executed population control programme that will reduce present growth rates must be undertaken.

Why the environmental crisis?

Environmental destruction in the Philippines is basically the result of too much pressure exerted on the natural resource base—forests and mineral and aquatic resources. Overexploitation of such resources by powerful business interests that tap them for quick profits has brought about serious ecological imbalances. These resources have also been overexploited to provide the export market with cheap raw materials. This has been to the detriment of our own local industries, which continue to depend on foreign goods.

Many of our environmental problems are the product of our

colonial history. Spain, Japan, and most importantly the US, have done considerable damage to our resource base during their reigns. Even at present, semi-colonial structures permit foreign business interests, such as those from the US and Japan, to exploit our resources.

The result is that access to resource utilisation is primarily in the hands of local ruling élites and foreign businessmen. Logging concessions in the hands of top businessmen and politicians control about 4.34 million hectares of our forests. Multinationals like Atlas, Benguet, Philiex and Marcopper control the country's mineral resources. Even in marine resources, huge trawlers dominate the coastal zones at the expense of small fishermen.

Our semi-colonial orientation has also enabled foreign interests to dictate technological choices to us which are unecological. The case of the Marcos Green Revolution, which has caused ecological havoc in agriculture, is a good example. Even at present, foreign consultants and World Bank-sponsored environmental projects are solicited by our own government to allegedly assist in resolving our environmental problems. Will they help? We are well aware of the experiences of other countries where these conservation programmes have been a flop.

The government and local and foreign business interests continue to blame the lowly *kaingineros*, the small fisherfolk, and the squatters for the Philippines' environmental problems. They cite figures which prove that these people consume a large portion of our resources and use technologies which are wasteful and degrading.

What is being missed is the fact that the unjust pattern of access to our natural resources is pushing these people to adopt subsistence-level technological systems. The invasion of the uplands by lowland farmers is a case in point. The failure of government agrarian reform programmes has been the major cause of the exhaustion of available land for small farmers, thus driving them to the uplands.

Worse, government measures for the protection of the environment have been characterised by graft and corruption. It is no secret that illegal loggers thrive because they enjoy the protection of top politicians, highly placed government officials, and military officers. Sometimes, the military are the illegal loggers themselves.

Solutions

If the government wishes to solve the environmental malady in a lasting way, it needs to direct its strategy towards addressing the problem of control over the country's natural resources. The great majority of our people, mostly peasants, workers and semi-workers, should have greater access to and control over our natural resources. The prerogative of tapping the country's riches now lies in the hands of foreign entities and local élites who view and treat these resources merely as sources of profit.

What is needed is a comprehensive environmental programme that should specifically attempt the following:

First, it should alter the unjust pattern of access based on wealth and political influence which now prevails. The bigger portion of our society should be able to utilise the country's resources as common owners, with all of them equitably benefiting from the proceeds. Present agrarian reform programmes should be overhauled so as to meet the demands of farmers for adequate access to land, capital and technology.

Second, the government should seriously embark on a programme to develop local and national industries. Our scarce raw materials should not go into the export market as cheap goods. Rather, local industries converting these raw materials to higher value-added goods for local consumption, and if necessary, for export, should be established. In this way, our resources would be utilised to meet the demands of our own people.

Third, the government, in tackling these problems, should make a point of involving communities in resource management and environmental issues. In this way, the government would ensure that strategies are attuned to the needs of the local populace, who have better answers to the problems. Dependence on foreign consultants will only cost the country more dollar drain and result in programmes that do no more than patch up problems.

My country may still recover from its environmental crisis. Unfortunately, it will be no simple task. It must be tackled by everyone. A people's environmental movement must be developed in order to collectively harness the people's energies towards confronting the crisis. And the time to act is now!

III

Modernisation
and
Culture

THE PRESSURE TO MODERNISE

Helena Norberg-Hodge

❖

Why do traditional societies break down upon their first sustained contact with the modern world?

The easy answer is that Western culture is intrinsically preferable—that blue jeans are simply better than homespun robes, the nuclear family better than the extended family.

My own view is very different. I believe that the most important reason for the breakdown of traditional cultures is the *psychological* pressure to modernise. I have come to this conclusion through almost two decades of close contact with the people of Ladakh, or 'Little Tibet'.

Much of the critique of conventional development has focused on the political and economic forces that foist modernisation on unprepared cultures, while the psychological side is largely neglected. And yet no one can deny the profound impact of glamourised Western images on the minds of young people who reject their own culture in favour of the 'American Dream'. Rambo and Barbie Dolls are making their way to the most remote corners of the world, with disastrous results.

This paper discusses some of the less obvious and insidious ways in which modernisation is carried to traditional cultures. I focus on the impact of the media, advertising and tourism, as well as the effects of Western-style education and economic models. Although most of my examples are drawn from Ladakh, virtually identical pressures are affecting people throughout the developing world.

The modern world comes to Ladakh
Ladakh is a high-altitude desert on the Tibetan Plateau in northernmost India. To all outward appearances, it is a wild and in-

hospitable place. In summer the land is parched and dry; in winter it is frozen solid by a fierce unrelenting cold. Harsh and barren, Ladakh's landforms have often been described as a 'moonscape'.

Almost nothing grows wild—not the smallest shrub, hardly a blade of grass. Even time seems to stand still, suspended on the thin air. Yet here, in one of the highest, dryest and coldest inhabited places on earth, the Ladakhis have for a thousand years not only survived, but prospered. Out of barren desert they have carved verdant oases—terraced fields of barley, wheat, apples, apricots and vegetables, irrigated with glacial meltwater brought many miles through stone-lined channels. Using little more than stone-age technologies and the scant resources at hand, the Ladakhis established a remarkably rich culture, one which met not only their material wants, but their psychological and spiritual needs as well.

Until 1962, Ladakh remained almost totally isolated from the forces of modernisation. In that year, however, in response to the conflict in Tibet, a road was built by the Indian Army to link the region with the rest of the country. With it came not only new consumer items and a government bureaucracy, but a first misleading impression of the world outside. Then, in 1975, the region was opened up to foreign tourists, and the process of 'development' began in earnest.

Speaking the language fluently from my first year in Ladakh, I have been able to observe almost as an insider the effect of these changes on the Ladakhis' perception of themselves. Within the space of a little more than a decade, feelings of pride gave way to what can best be described as a 'cultural inferiority complex'. In the modern sector today, most young Ladakhis—the teenage boys in particular—are ashamed of their cultural roots and desperate to appear modern.

Tourism

When tourism first began in Ladakh, it was as though people from another planet suddenly descended on the region. Looking at the modern world from something of a Ladakhi perspective, I became aware how much more successful our culture looks from the outside than we experience it on the inside.

Each day many tourists would spend as much as a hundred dollars—an amount roughly equivalent to someone spending fifty thousand dollars a day in America. In the traditional subsistence economy, money played a minor role, and was used primarily for luxuries—jewelry, silver, and gold. Basic needs— food, clothing and shelter—were provided for without money. The labour one needed was free of charge, part of an intricate web of human relationships.

Ladakhis did not realise that money played a completely different role for the foreigners, that back home they needed it to survive, that food, clothing and shelter all cost money—a lot of money. Compared to these strangers, the Ladakhis suddenly felt poor.

This new attitude contrasted dramatically with the Ladakhis' earlier self-confidence. In 1975, I was shown around the remote village of Hemis Shukpachan by a young Ladakhi named Tsewang. It seemed to me that all the houses we saw were especially large and beautiful. I asked Tsewang to show me the houses where the poor people lived. Tsewang looked perplexed a moment, then responded, 'We don't have any poor people here.'

Eight years later I overheard Tsewang talking to some tourists. 'If you could only help us Ladakhis,' he was saying, 'we're so poor.'

Besides giving the illusion that all Westerners are multi-millionaires, tourism and Western media images also help perpetuate another myth about modern life—that we never work. It looks as though our technologies do the work for us. In industrial society today, we actually spend *more* hours working than people in rural, agrarian economies. But that is not how it looks to the Ladakhis. For them, work is physical work: ploughing, walking, carrying things. A person sitting behind the wheel of a car or pushing buttons on a typewriter doesn't appear to be working.

One day I spent ten hours writing letters. I was exhausted, stressed, and had a headache. That evening, when I complained about being tired because of having worked so hard, the Ladakhi family I was staying with laughed; they thought I was joking. In their eyes I had not been working. I had been sitting in front of a table, nice and clean, no sweat on my brow, pushing a pen across a piece of paper. This was not work.

Media images

Development has brought not only tourism, but also Western and Indian films and, more recently, television. Together they provide overwhelming images of luxury and power. There are countless tools and magical gadgets. And there are machines— machines to take pictures, machines to tell the time, machines to make fire, to travel from one place to another, to talk with someone far away. Machines can do everything for you; it's no wonder the tourists look so clean and have such soft, white hands.

Media images focus on the rich, the beautiful, and the brave, whose lives are endless action and glamour. For young Ladakhis, the picture is irresistible. It is an overwhelmingly exciting version of an urban 'American Dream', with an emphasis on speed, youthfulness, super-cleanliness, beauty, fashion and competitiveness. 'Progress' is also stressed: humans dominate nature, while technological change is embraced at all costs.

In contrast to these utopian images from another culture, village life seems primitive, silly and inefficient. The one-dimensional view of modern life becomes a slap in the face. Young Ladakhis—who are asked by their parents to choose a way of life that involves working in the fields and getting their hands dirty for very little or no money—feel ashamed of their own culture. Traditional Ladakh seems absurd compared with the world of the tourists and film heroes.

This same pattern is being repeated in rural areas all over the South, where millions of young people believe modern Western culture to be far superior to their own. This is not surprising: looking as they do from the outside, all they can see is the material side of the modern world—the side in which Western culture excels. They cannot so readily see the social or psychological dimensions—the stress, the loneliness, the fear of growing old. Nor can they see environmental decay, inflation, or unemployment. On the other hand, they know their own culture inside out, including all its limitations and imperfections.

In Ladakh and elsewhere in the South, the sudden influx of Western influences has caused some people—particularly the young men—to develop feelings of inferiority. They reject their own culture wholesale, and at the same time eagerly embrace the new one. They rush after the symbols of modernity: sunglasses, Walkmans and blue jeans—not because they find those jeans more attractive or comfortable, but because they are symbols of modern life.

Modern symbols have also contributed to an increase in aggression in Ladakh. Young boys now see violence glamourised on the screen. From Western-style films, they can easily get the impression that if they want to be modern, they should smoke one cigarette after another, get a fast car, and race through the countryside shooting people left and right!

It has been painful for me to see the changes in young Ladakhi friends. Of course they don't all turn violent, but they do become angry and less secure. I have seen a gentle culture change—a culture in which men, even young men, were not in the slightest bit ashamed to cuddle a baby or to be loving and soft with their grandmothers.

Western-style education

No one can deny the value of real education—the widening and enrichment of knowledge. But today in the Third World, education has become something quite different. It isolates children from their culture and from nature, training them instead to become narrow specialists in a Westernised urban environment. This process has been particularly striking in Ladakh, where modern schooling acts almost as a blindfold, preventing children from seeing the very context in which they live. They leave school unable to use their own resources, unable to function in their own world.

With the exception of religious training in the monasteries, Ladakh's traditional culture had no separate process called 'education'. Instead, education was the product of a person's intimate relationship with their community and their ecosystem. Children learned from grandparents, family and friends, and from the natural world.

Helping with the sowing, for instance, they would learn that on one side of the village it was a little warmer, on the other side a little colder. From their own experience children would come to distinguish between different strains of barley and the specific growing conditions each strain preferred. They learned to recognise even the tiniest wild plant and how to use it, and how to pick out a particular animal on a faraway mountain slope. They learned about connection, process, and change, about the intricate web of fluctuating relationships in the natural world around them.

For generation after generation, Ladakhis grew up learning how to provide themselves with clothing and shelter; how to make shoes out of yak skin and robes from the wool of sheep; how to build houses out of mud and stone. Education was location-specific and nurtured an intimate relationship with the living world. It gave children an intuitive awareness that allowed them, as they grew older, to use resources in an effective and sustainable way.

None of that knowledge is provided in the modern school. Children are trained to become specialists in a technological, rather than an ecological, society. School is a place to forget traditional skills, and worse, to look down on them.

Western education first came to Ladakhi villages in the 1970s. Today there are about two hundred schools. The basic curriculum is a poor imitation of that taught in other parts of India, which itself is an imitation of British education. There is almost nothing Ladakhi about it.

Once, while visiting a classroom in the capital, Leh, I saw a drawing in a textbook of a child's bedroom that could have been in London or New York. It showed a pile of neatly folded handkerchiefs on a four-poster bed and gave instructions as to which drawer of the vanity unit to keep them in. Many other schoolbooks were equally absurd and inappropriate. For homework in one class, pupils were supposed to figure out the angle of incidence that the Leaning Tower of Pisa makes with the ground. Another time they were struggling with an English translation of *The Iliad*.

Most of the skills Ladakhi children learn in school will never be of real use to them. In essence, they receive an inferior version of an education appropriate for a New Yorker. They learn from books written by people who have never set foot in Ladakh, who know nothing about growing barley at 12,000 feet or about making houses out of sun-dried bricks.

This situation is not unique to Ladakh. In every corner of the world today, the process called 'education' is based on the same assumptions and the same Eurocentric model. The focus is on faraway facts and figures, on 'universal' knowledge. The books propagate information that is meant to be appropriate for the entire planet. But since only a kind of knowledge that is far removed from specific ecosystems and cultures can be universally applicable, what children learn is essentially synthetic, divorced from the living context. If they go on to higher education, they may learn about building houses, but these houses will be of concrete and steel, the universal box. So too, if they study agriculture, they will learn about industrial farming: chemical fertilisers and pesticides, large machinery and hybrid seeds. The Western educational system is making us all poorer by teaching people around the world to use the same industrial resources, ignoring those of their own environment. In this way education is creating artificial scarcity and inducing competition.

In Ladakh and elsewhere, modern education not only ignores local resources, but worse still, robs children of their self-esteem. Everything in school promotes the Western model and, as a direct consequence, makes children think of themselves and their traditions as inferior.

A few years ago, Ladakhi schoolchildren were asked to imagine their region in the year 2000. A little girl wrote, 'Before 1974, Ladakh was not known to the world. People were uncivilised. There was a smile on every face. They don't need money. Whatever they had was enough for them.' In another essay a child wrote, 'They sing their own songs like they feel disgrace, but they sing English and Hindi songs with great interest... But in these days we find that maximum people and persons didn't wear our own dress, like feeling disgrace.'

Education pulls people away from agriculture into the city, where they become dependent on the money economy. Traditionally there was no such thing as unemployment. But in the modern sector there is now intense competition for a very limited number of paying jobs, principally in the government. As a result, unemployment is already a serious problem.

Modern education has brought some obvious benefits, like improvement in the literacy rate. It has also enabled the Ladakhis to be more informed about the forces at play in the world outside. In so doing, however, it has divided Ladakhis from each other and the land and put them on the lowest rung of the global economic ladder.

Local economy vs. global economy
In the past individual Ladakhis had real power, since political and economic units were small and each person was able to deal directly with the other members of the community. Today, 'development' is hooking people into ever-larger units. In political terms, each Ladakhi has become one of 800 million, and, as part of the global economy, one of several billion.

In the traditional economy, everyone knew they had to depend directly on family, friends and neighbours. But in the new economic system, one's political and economic interactions take a detour via an anonymous bureaucracy. The fabric of local interdependence is disintegrating as the distance between people increases.

So too are traditional levels of tolerance and cooperation. This is particularly true in the villages near Leh, where disputes and acrimony within close-knit communities and even families have dramatically increased in the last few years. I have even seen heated arguments over the allocation of irrigation water, a procedure that had previously been managed smoothly within a cooperative framework.

As mutual aid is replaced by a dependence on faraway forces, people begin to feel powerless to make decisions over their own lives. At all levels passivity, even apathy, is setting in; people are abdicating personal responsibility. In the traditional village, for example, repairing irrigation canals was a task shared by the whole community. As soon as a channel developed a leak, groups of people would start working away with shovels patching it up. Now people see this as the government's responsibility, and will let a channel go on leaking until the job is done for them. The more the government does for the villagers, the less they feel inclined to help themselves.

In the process, Ladakhis are starting to change their perception of the past. In the early days, people would tell me there had never been hunger in Ladakh. I kept hearing the expression *tungbos zabos*: 'enough to drink, enough to eat'. Now, particularly in the modern sector, people can be heard saying, 'Development is essential; in the past we couldn't manage, we didn't have enough.'

The cultural centralisation that occurs through the media is also contributing to this passivity, as well as to a growing insecurity. Traditionally, village life included lots of dancing, singing and theatre. People of all ages joined in. In a group sitting around the fire, even toddlers would dance, with the help of older siblings or friends. Everyone knew how to sing, to act, to play music. Now that the radio has come to Ladakh, people do not need to sing their own songs or tell their own stories. Instead, they can sit and listen to the best singer, the best storyteller. But the result is that people become inhibited and self-conscious. They are no longer comparing themselves to neighbours and friends, who are real people—some better at singing but perhaps not so good at dancing—and one is never as good as the stars on the radio. Community ties are also broken when people sit passively listening to the very best rather than making music or dancing together.

Artificial needs
Before the changes brought by tourism and modernisation, the Ladakhis were self-sufficient, psychologically as well as materially.

There was no desire for the sort of development that later came to be seen as a 'need'. Time and again, when I asked people about the changes that were coming they showed no great interest in being modernised; sometimes they were even suspicious. In remote areas, when a road was about to be built, people at best felt ambivalent about the prospect. The same was true of electricity. I remember distinctly how, in 1975, people in Stagmo village laughed about the fuss that was being made to bring electric lights to neighbouring villages. They thought it was a joke that so much effort and money was spent on what they took to be a ludicrous gain: 'Is it worth all that bother just to have that thing dangling from your ceiling?'

Two years ago, when I arrived in the same village to meet the council, the first thing they said to me was, 'Why do you bother to come to our backward village where we live in the dark?' They said it jokingly, but it was obvious they were ashamed of the fact they did not have electricity.

Before people's sense of self-respect and self-worth had been shaken, they did not need electricity to prove they were civilised. But within a short period the forces of development so undermined people's self-esteem that not only electricity, but Punjabi rice and plastic have become needs. I have seen people proudly wear wristwatches they cannot read and for which they have no use. And as the desire to appear modern grows, people are rejecting their own culture. Even the traditional foods are no longer a source of pride. Now when I'm a guest in a village, people apologise if they serve *ngamphe* instead of instant noodles.

Surprisingly, perhaps, modernisation in Ladakh is also leading to a loss of individuality. As people become self-conscious and insecure, they feel pressure to conform, to live up to the idealised images—to the American Dream. By contrast, in the traditional village, where everyone wears the same clothes and looks the same to the causal observer, there seems to be more freedom to relax and be who you really are. As part of a close-knit community, people feel secure enough to be themselves.

As local economic and political ties are broken, the people around you become more and more anonymous. At the same time, life speeds up and mobility increases, making even familiar relations more superficial and brief. The connections between people are reduced largely to externals. A person comes to be identified with what they have rather than what they are, and disappear behind their clothes and other belongings.

A people divided

Perhaps the most tragic of all the changes I have observed in Ladakh is the vicious circle in which individual insecurity contributes to a weakening of family and community ties, which in turn further shakes individual self-esteem. Consumerism plays a central role in this whole process, since emotional insecurity contributes to a hunger for material status symbols. The need for recognition and acceptance fuels the drive to acquire possessions—possessions that will make you somebody. Ultimately, this is a far more important motivating force than a fascination for the things themselves.

It is heartbreaking to see people buying things to be admired, respected, and ultimately loved, when in fact the effect is almost inevitably the opposite. The individual with the new shiny car is set apart, and this furthers the need to be accepted. A cycle is set in motion in which people become more and more divided from themselves and from one another.

I've seen people divided from one another in many ways. A gap is developing between young and old, male and female, rich and poor, Buddhist and Muslim. The newly created division between modern, educated expert and illiterate, 'backward' farmer is perhaps the biggest of all. Modernised inhabitants of Leh have more in common with someone from Delhi or Calcutta than with their own relatives who have remained on the land, and they tend to look down on anyone less modern. Some children living in the modern sector are now so distanced from their parents and grandparents that they don't even speak the same language. Educated in Urdu and English, they are losing mastery of their native tongue.

Around the world, another consequence of development is that the men leave their families in the rural sector to earn money in the modern economy. The men become part of the techno-logically-based life outside the home and are seen as the only productive members of society. In Ladakh, the roles of male and female are becoming increasingly polarised as their work becomes more differentiated.

Women become invisible shadows. They do not earn money for their work, so they are no longer seen as 'productive'. Their work is not included as part of the Gross National Product. In government statistics, the 10% or so of Ladakhis who work in the modern sector are listed according to their occupations; the other 90%—housewives and traditional farmers—are lumped together as 'non-workers'. Farmers and women are coming to be viewed as inferior, and they themselves are developing feelings of insecurity and inadequacy.

Over the years I have seen the strong, outgoing women of Ladakh being replaced by a new generation—women who are unsure of themselves and extremely concerned with their appearance. Traditionally, the way a woman looked was important, but her capabilities—including tolerance and social skills—were much more appreciated.

Despite their new dominant role, men also clearly suffer as a result of the breakdown of family and community ties. Among other things, they are deprived of contact with children. When they are young, the new macho image prevents them from showing any affection, while in later life as fathers, their work keeps them away from home.

Breaking the bonds between young and old

In the traditional culture children benefited not only from continuous contact with both mother and father, but also from a way of life in which different age groups constantly interacted. It was quite natural for older children to feel a sense of responsibility for the younger ones. A younger child in turn looked up to the older ones with respect and admiration, and sought to be like them. Growing up was a natural, non-competitive learning process.

Now children are split into different age groups at school.

This sort of leveling has a very destructive effect. By artificially creating social units in which everyone is the same age, the ability of children to help and to learn from each other is greatly reduced. Instead, conditions for competition are automatically created, because each child is put under pressure to be just as good as the next one. In a group of ten children of quite different ages, there will naturally be much more cooperation than in a group of ten twelve-year-olds.

The division into different age groups is not limited to school. Now there is a tendency to spend time exclusively with one's peers. As a result, a mutual intolerance between young and old has emerged. Young children nowadays have less and less contact with their grandparents, who often remain behind in the village. Living with many traditional families over the years, I have witnessed the depth of the bond between children and their grandparents. It is clearly a natural relationship, with a very different dimension from that between parent and child. To sever this connection is a profound tragedy.

Similar pressures contribute to the breakdown of the traditional family. The Western model of the nuclear family is now seen as the norm, and Ladakhis are beginning to feel ashamed about their traditional practice of polyandry, one of the cultural controls on population growth. As young people reject the old family structure in favour of monogamy, the population is rising significantly. At the same time, monastic life is losing its status, and the number of celibate monks and nuns is decreasing. This too contributes to population increase.

Ethnic conflict

Interestingly, a number of Ladakhis have linked the rise in birth rates to the advent of modern democracy. 'Power is a question of votes' is a current slogan, meaning that in the modern sector, the larger your group, the greater your access to power. Competition for jobs and political representation within the new centralised structures is increasingly dividing Ladakhis. Ethnic and religious differences have taken on a political dimension, causing bitterness and envy on a scale hitherto unknown.

This new rivalry is one of the most painful divisions that I have seen in Ladakh. Ironically, it has grown in proportion to the decline of traditional religious devotion. When I first arrived, I was struck by the mutual respect and cooperation between Buddhists and Muslims. But within the last few years, growing competition has actually culminated in violence. Earlier there had been individual cases of friction, but the first time I noticed any signs of group tension was in 1986, when I heard Ladakhi friends starting to define people according to whether they were Buddhist or Muslim. In the following years, there were signs here and there that all was not well, but no one was prepared for what happened in the summer of 1989, when fighting suddenly broke out between the two groups. There were major disturbances in Leh bazaar, four people were shot dead by police, and much of Ladakh was placed under curfew.

Since then, open confrontation has died down, but mistrust and prejudice on both sides continue to mar relations. For a people unused to violence and discord, this has been a traumatic experience. One Muslim woman could have been speaking for all Ladakhis when she tearfully told me, 'These events have torn my family apart. Some of them are Buddhists, some are Muslims, and now they are not even speaking to each other.'

The immediate cause of the disturbances was the growing perception among the Buddhists that the Muslim-dominated stated government was discriminating against them in favour of the local Muslim population. The Muslims for their part were becoming anxious that as a minority group they had to defend their interests in the face of political assertiveness by the Buddhist majority.

However, the underlying reasons for the violence are much more far-reaching. What is happening in Ladakh is not an isolated phenomenon. The tensions between the Muslims of Kashmir and the Hindu-dominated central government in Delhi, the Hindus and the Buddhist government in Bhutan, and the Buddhists and the Hindu government in Nepal, along with countless similar disturbances around the world, are, I believe, all connected to the same underlying cause. The present development model is intensely centralising, pulling diverse peoples from rural areas into large urban centres and placing power and decision-making in the hands of a few. In these centres, job opportunities are scarce, community ties are broken, and competition increases dramatically. Young men in particular, who have been educated for jobs in the modern sector, find themselves engaged in a competitive struggle for survival. In this situation, any religious or ethnic differences quite naturally become exaggerated and distorted. In addition, the group in power inevitably has a tendency to favour its own kind, while the rest often suffer discrimination.

Most people believe that ethnic conflict is an inevitable consequence of differing cultural and religious traditions. In the South, there is an awareness that modernisation is exacerbating tensions; but people generally conclude that this is a temporary phase on the road to 'progress', a phase that will only end once development has erased cultural differences and created a totally secular society. On the other hand, Westerners attribute overt religious and ethnic strife to the liberating influence of democracy. Conflict, they assume, always smouldered beneath the surface, and only government repression kept it from bursting into flame.

It is easy to understand why people lay the blame at the feet of tradition rather than modernity. Certainly ethnic friction is a phenomenon which predates colonialism and modernisation. But after nearly two decades of firsthand experience on the Indian subcontinent, I am convinced that 'development' not only exacerbates tensions but in many cases actually creates them. As I have pointed out, development causes artificial scarcity, which inevitably leads to greater competition. Just as importantly, it puts pressure on people to conform to a standard Western ideal—blonde, blue-eyed, beautiful and rich—that is impossibly out of reach.

Striving for such an ideal means rejecting one's own culture and roots—in effect, denying one's own identity. The inevitable result is alienation, resentment and anger. I am convinced that much of the violence and fundamentalism in the world today is a product of this process. In the industrialised world we are becoming increasingly aware of the impact of glamorous media and advertising images on individual self-esteem—resulting in problems that range from eating disorders like anorexia and bulimia, to violence over high-priced and 'prestigious' sneakers and other articles of clothing. In the South, where the gulf between reality and the Western ideal is so much wider, the psychological impacts are that much more severe.

The old culture and the new
There were many real problems in the traditional society and development does bring some real improvements. However, when one examines the fundamentally important relation-ships—to the land, to other people, and to oneself—develop-ment takes on a different light. Viewed from this perspective, the differences between the old and the new become stark and disturbing. It becomes clear that the traditional nature-based society, with all its flaws and limitations, was more sustainable, both socially and environmentally. It was the result of a dialogue between human beings and their surroundings, a continuing coevolution that meant that—during two thousand years of trial and error—the culture kept changing. Ladakh's traditional Buddhist worldview emphasised change, but change within a framework of compassion and a profound understanding of the interconnectedness of all phenomena.

The old culture reflected fundamental human needs while respecting natural limits. And it worked. It worked for nature, and it worked for people. The various connecting relationships in the traditional system were mutually reinforcing, and encour-aged harmony and stability. Most importantly of all, having seen my friends change so dramatically, I have no doubt that the bonds and responsibilities of the traditional society, far from being a burden, offered a profound sense of security, which seems to be a prerequisite for inner peace and contentedness. I am convinced that people were significantly happier before development than they are today. And what criteria for judging a society could be more important: in social terms, the well-being of the people; in environmental terms, sustainability.

By comparison, the new Ladakh scores very poorly when judged by these criteria. The modern culture is producing an array of environmental problems that, if unchecked, will lead to irreversible decline; socially, it is leading to the breakdown of community and the undermining of personal identity.

At my lectures in Europe and North America, people often ask the same question. Having seen pictures of the wide uninhibited smiles of the Ladakhis and the beauty of the traditional art, architecture and landscape contrasted with the meanness and spiritual poverty of the modern sector, they say, 'How can the Ladakhis possibly want to give up their traditional way of life? They must want the change, there must have been some flaw in the traditional culture that makes them want to abandon it. It can't have been that good.'

It is easy to understand why people make such assumptions. Had I not spoken the language fluently in my first year in Ladakh, had I not been lucky enough to live closely with the Ladakhi people, I would almost certainly thought the same way. But the Ladakhis I lived with were content; they were not dissatisfied with their lives. I remember how shocked they used to be when I told them that in my country, many people were so unhappy that they had to see a doctor. Their mouths would drop open, and they would stare in disbelief. It was beyond their experience. A sense of deep-rooted contentedness was something they took for granted.

If the Ladakhis had been eager to adopt another culture, they could easily have done so. Leh was for centuries a centre of trans-Asian trade. The Ladakhis themselves traveled both as pilgrims and traders, and were exposed to a variety of foreign influences. In many instances they absorbed the materials and practices of other cultures, and used them to enhance their own. But it was never a question of adopting another culture wholesale. If someone from China came to Leh, the result was not that the young suddenly wanted to put on Chinese hats, eat only Chinese food, and speak the Chinese language.

As I have tried to show, the pressures that lead to the breakdown of a culture are many and varied. But the most important elements have to do with the psychological pressures that create a sense of cultural inferiority, and the fact that people cannot have an overview of what is happening to them as they stand in the middle of the development process. Modernisation is not perceived as a threat to the culture. Each change that comes along usually looks like an unconditional improvement; there is no way of anticipating the negative long-term consequences. Since people have almost no information about the impact development has had in other parts of the world, it is only in looking back that any destructive effects become obvious.

By now, most Ladakhis deem development necessary. And although the traditional society compares so favourably with the new, it was of course not perfect; there was certainly room for improvement.

But does development have to mean destruction? I do not believe so. I am convinced that the Ladakhis and other traditional peoples can raise their standard of living without sacrificing the sort of social and ecological balance that they have enjoyed for centuries. To do so, however, they would need to maintain their self-respect and self-reliance. They would need to build on their own ancient foundations rather than tearing them down, as is the way of conventional development.

This paper is based on excerpts from Helena Norberg-Hodge's book, Ancient Futures: Learning from Ladakh *(Sierra Club Books, San Francisco, and Rider Books, London, 1991).*

THE THIRD WORLD:
A Crisis of Development

S. M. Mohamed Idris

❖

Today, after many years of political independence, after so-called 'Development Decades', after all the efforts of United Nations agencies, after the hundreds of pious declarations on aid, trade and development, and after the millions and millions of pages printed and published on development—after all these phenomena, most of the people in the Third World continue to be poor, unemployed, and homeless; and at this very moment, millions of children are starving to death.

The Consumers' Association of Penang (CAP) in Malaysia has tried to set an example amongst Third World countries by tackling the issues of pollution and environmental deterioration, and by looking for solutions to the problems of meeting basic human needs (such as food, nutrition, and health) and business malpractices.

In the course of carrying out these efforts, some people have asked us what the alternatives are for the future. What sort of development can we have which will not destroy the environment or damage our health, which does not waste resources, which protects consumers from exploitation, which provides for the basic needs of ordinary people, and which at the same time results in human happiness and peace of mind (instead of all the mental stresses, hypertension and alienation which modern industrialised people seem to suffer)?

Clearly, these questions are framed in the light of bitter, first-hand experience of the negative impact of the Western development model on Third World societies.

The Western Monoculture

In the social and cultural spheres, the industrial world seems to have a great hold over the Third World. Third World countries have consciously or unconsciously imported models of education, communications, health care, housing, and transportation from the industrial countries. Most of these models are profoundly unsuitable and inappropriate for solving the problems of the majority of people in the Third World. Instead, these models have mainly benefited a small élite. For instance, billions of dollars are spent on imported motor cars and billions more on building roads and bridges, while public transport systems remain inadequate in most Third World countries.

The minds and motivations of Third World people are also increasingly influenced by the media and fashion industries of the industrial countries. As television programmes, films, videos, records, books and magazines produced in the industrial countries invade every nook and corner of the Third World, the culture and ways of life of the people and the community are disrupted. Traditional community dances give way to disco and break-dance, and traditional musical forms surrender to the beat of Michael Jackson or his local imitators. A large section of Third World society—from the business and middle class right down to the worker and farmer—excitedly watch the latest antics on 'Dallas' and 'Dynasty', or on the dozens of Hong Kong versions of these American programmes. Coffee shop conversations in many Third World countries are often dominated by the latest developments in the Hollywood film star circuit, or the newest contest between sportsmen and sportswomen in Wimbledon or Wembley.

The marketing of status and style

The consumer culture of the North now pervades almost all aspects of life in the South. This culture is in reality a way of thinking and a way of life generated by advertisements, cinema shows, pop songs, magazines, comics, and other channels of the mass media. The objective of the consumer culture is to persuade the consumer to buy more and more products, whether these products are good for her or not.

As a result of this bombardment, the consumer is made to feel insecure unless she smokes a cigarette, unless he buys a certain brand of haircream, unless she uses a certain brand of lipstick, unless they change motorcars once every two years, unless the colour of the curtains at home matches the colour of the carpet.

Status and style are being attached to such products as cigarettes, cosmetics, soft drinks, artificial baby food, and fashionable clothes, and such activities as disco dancing and motorbike racing. A large part of the already meagre salary of the lower-income groups is being siphoned off to such worthless goods and activities, leaving little money and time for food and healthy recreation. Schoolchildren prefer to spend their pocket money on sweets and soft drinks rather than on wholesome food. And, tragically, many mothers have given up breast-feeding due to the false association of the bottle with status. The poor baby is the one to suffer.

A whole new generation of people has been brought up with the idea that their main aim in life should be grabbing for themselves as many things as possible. Thus, people want to possess more and better things than their neighbours or their friends because this gives them higher status and prestige. As rational people, we should not fall into the trap of this consumer culture for it is often harmful; it generates egoism, individualism and rivalry between individuals who measure their worth according to the size of their house, the make of their car, the beauty of their wives or the wealth of their husbands, and the intelligence of their children.

The mass media

Information determines how we think, how we view ourselves and how we view others. Throughout history, information has been the special privilege of the chosen few. In ancient times, it was the scribes and the priestly class who controlled information. Very often it was shrouded in myth and secrecy and in a language that the common people did not understand. In this manner, information and access to it was controlled and excluded from the vast majority. Inevitably it was used as a tool to mislead and subjugate people.

In our modern times, information systems are no better. They are now very sophisticated, neatly-packaged but complicated, and controlled by the multinational corporations. For example, it is an undisputed fact that over 90 percent of the foreign news generated by the world's newspapers is provided by just four Western news agencies: United Press International (UPI), Associated Press (AP), Reuters, and Agence France Presse (AFP). Similarly, TV programmes beamed into all Third World homes are heavily Western and American. Information systems are now controlled by the mass media, which is both oligopolistic and alien.

The immense power of the mass media has enabled it to mould beliefs, attitudes, values and lifestyles. It has become the purveyor of a global culture, a culture based on a pattern of consumption which is compulsive and devoid of meaning, a culture which is the same everywhere in food, dress, song, leisure and outlook.

This consumer culture has not only managed to destroy what good there is in our indigenous cultures and societies, it has alienated our young, robbing them of their confidence and dignity. This form of cultural domination distorts the personality and engenders an inferiority complex in our people.

All Third World societies have their own indigenous systems of communication through social activities, songs and folklore. But modern mass media has now taken over this function. Often the only information—the only form of truth—is provided by this media. People then tend to believe the information fed by the mass media and the messages conveyed by it. As a result, they discard good wholesome nutrition for fast foods promoted by advertisers. They abandon traditional skills in household arts and crafts, because they have come to believe that modern plastic furniture, mats and utensils are far superior.

In our societies—with immense poverty, illiteracy and superstition, with vast differences separating the urban rich and the rural poor—people are told that their problems are due to population explosion, traditionalism and indolence. The root causes, the *real* reasons for their poverty, malnutrition and deprivation are not explained. They are not linked to modern lifestyles, economic structures and social inequalities. Too often, it is not in the interest of the people in power to disseminate accurate and truthful information. In this climate, true information is denied and manipulated, and the average citizen is misled by misinformation and disinformation.

Modern addictions
The stress and alienation of the Western lifestyle, combined with pressure from the mass media, leads people to use addicting and harmful substances. Addiction to drugs, tobacco and alcohol is a great social menace, and is sapping the energy and destroying the health of our youth. In 1978, there were an estimated 300,000 to 400,000 drug addicts in Malaysia. An increasing proportion of 'respectable' adults are dependent on tranquilisers, pep pills and sleeping pills, many of which have dangerous side effects. Millions of Malaysians smoke their way to lung cancer, heart attacks and other tobacco-related ailments. Alcohol addiction also takes its toll, whether in deaths caused by illegal *samsu*, or liver destruction caused by the higher-class spirits.

The consumption of sugar and other sweetened products has also increased over the years. The reason for this is the reckless manner in which people have abandoned themselves to the manipulations of manufacturers. It is not pointed out that sugar is a very dangerous substance, that it has no benefits and instead contributes to a lot of health problems such as dental caries, diabetes, obesity and skin problems. All of these problems are very prevalent in Malaysia.

The World Economic and Environmental Crisis

The world faces many severe and fundamental problems which undermine the assumptions which have long been held about development. The world's natural resources are being depleted so fast that many scientists now doubt if the present development strategy, with its emphasis on capital-intensive methods, can continue for much longer.

Economists and planners have usually focused on the flow of income as measured in the Gross National Product (GNP). But they have forgotten to take into account the stock of resources which gives rise to and makes possible this flow of income. This stock of resources has been used in the years since the Industrial Revolution to enable rapid economic growth. But with the depletion of these resources, the future flow of income—the future of the GNP itself—is now being threatened.

Indeed, the rapid depletion of the world's natural resources is the most important factor which will determine the nature and the shape of human society in the years ahead. Yet this central fact of life has yet to be fully taken account of by governments, planners, academics, economists or businessmen, including those in Malaysia. Development plans are still being drawn up on the assumption that all the energy, mineral, and biological resources that we have used in the past and are using in the present will always be with us.

The truth is that man has already overexploited nature, and in a few years there will be nothing left to exploit. By the year 2000, for instance, the tropical forests will largely have disappeared from the face of the Earth, taking along with them millions of species of animal and plant life which have existed for hundreds of millions of years.

The results of this rapid resource depletion are already being seen in the present world economic recession and the rate of inflation in the world. This economic crisis has been spurred on by a shortage of basic materials and higher costs of production. For instance, as good quality tin ore near the ground surface is depleted, production costs go higher and higher as the miners have to dig longer and deeper to obtain tin of lower grade. The rising cost of production eventually pushes the price up, although at the moment the world recession is having a depressive effect on commodity prices.

Economic crisis in the Third World

Developing countries like Malaysia are being badly affected by world recession and inflation. A large part of our economic growth has depended on the continuous exploitation and export of resources such as timber, tin and petroleum. These resources are going to be exhausted in the near future—timber and petroleum in the 1990s while tin production is already dropping annually and may be depleted in 30 or 40 years.

Third World countries have become even more dependent on Northern countries in the post-World War II period, and more of our resources and labour go into commodities exported to the rich countries. Trade policies are biased toward the industrial nations, thus causing hundreds of billions of dollars of real resources to be transferred from poor countries to rich countries. Prices for Malaysia's commodities—like rubber, palm oil, pepper, cocoa and tin—are being depressed by the low demand from the rest of the world. More than a quarter of the workforce in the Malaysian tin industry has been retrenched due to the collapse of tin prices, and three million people depending on rubber for a source of income have suffered a drop in real income due to the declining price of rubber.

In the Third World, the best quality lands are taken up by export crops. The richest of our forest, mineral and metal resources are exported. Our best brains and a very substantial part of our labour force are used in the service of transnational corporations owned by the rich countries. Almost all our traded goods are carried on ships owned by rich countries. The international chain of commodity traders, wholesalers and retailers are controlled by the same countries. And finally, our top researchers spend their long hours conducting research for institutions ultimately controlled by the administrations of the rich countries, and large numbers of our academics, doctors and scientists migrate to their shores. All these indisputable facts have led to a continuous drain of money and resources from the Third World.

During the colonial era, the colonial masters squeezed wealth from their colonies to develop their own countries. Today, this is even more true. The belief that rich countries are giving 'development aid' to poor countries is essentially a myth. In reality, the Third World is channeling resources and funds to the industrial countries, in the form of profits on investment, interest on loans, royalties for technology, fees for management and consultancy, losses due to terms-of-trade decline, and taxes lost through transfer pricing by transnational corporations.

When Third World governments try to break away from the economic or social chains that bind them to the industrial nations, they are often blocked. For instance, when Bangladesh decided to ban dangerous or worthless pharmaceutical drugs, the US government intervened on behalf of the drug industry. And when Third World countries do not follow policies that please the major powers, they can be threatened with invasion, as happened when the US invaded Grenada. When international agencies like UNESCO or the ILO or UNCTAD endeavour to take up issues on behalf of the Third World, they are threatened with a pullout of funds, as the US has done.

Reevaluating development policies

The purpose of development should be to increase the well-being of individuals throughout society. With more than a decade of experience analysing current development patterns, we have come to realise that the individual's problems and welfare are linked to the conditions of his or her community. For instance, a housewife finds that fish prices have shot up and there is a shortage of fish in the market. This is linked to the plight of the fishing community, whose fish stocks are being depleted by trawler boats introduced by a development project.

We have also come to the conclusion that the problems of local communities are related to policies at the national level. For instance, land and housing shortages, the rising cost of living, and the lack of proper sanitation are all experienced at the local community level. But their roots lie in the absence of proper national planning, or worse, of wrong national planning, leading to top-heavy, élite-oriented government bureaucracies.

And finally, we have come to realise that problems at the

national level cannot be discussed separately from conditions prevailing at the international level. For instance, as mentioned previously, a large segment of our society has been adversely affected by the world recession and falling commodity prices.

Because of the current pattern of development, the needs and welfare of the individual consumer, the economic and social situation of his and her community, and of the nation as a whole, are all tied up with and influenced by events and developments at the world level. And this is true whether we are talking of economics, social issues, or cultural ways of life. It is at times like these, when our dependence on the world economy is so acutely felt, that we realise the need for a self-reliant type of development, in which we produce sufficient quantities of food and basic consumer commodities for use by the local population, rather than exporting raw materials in exchange for manufactured items and food.

Planting an indigenous seed

It is not only in the area of production or trade that we need to be self-reliant. We also need to develop our own ideas and policies on development to suit local needs and to meet local aspirations. For far too long we have borrowed and put into practice theories of growth and development which were written in the West by so-called experts, many of whom have never lived in or even visited developing countries. Development theories and plans must rise up from indigenous soil itself, and not be plucked out from European or American skies and implanted. A wrong seed will not grow well in Third World countries.

One worrying feature of our development is that we seem to have been too much taken up with the belief that modernisation *per se* will provide the key to our development, a belief which originated from the West and was passed on to us, even though in the West itself this assumption is being abandoned. In developing our countries, we seem to think that big and prestigious projects are the hallmark of modernisation and growth.

But in the light of the resource depletion problem and the economic recession, it may be necessary at this stage to reevaluate development projects which are either existing or are being planned. Such a reevaluation is necessary to ensure that a huge waste of funds does not take place. For instance, current plans to build new highways worth billions of dollars may be economically unwise given the accelerating prices of oil and motor vehicles, and the expected decline in motorcar usage in the years to come.

Projects such as billion-dollar highways and multi-storey office blocks may seem impressive from the point of view of modernisation, but they may well be obsolete and useless in the very near future. Projects which do not meet stringent requirements of efficient resource use, environmental impact and social usefulness should be reconsidered or should not be approved. On the other hand, projects which are efficient in their use of resources, which are ecologically sound and which meet the basic material and spiritual needs of the majority of people should be actively sought out and promoted.

One of the main activities of CAP is helping rural communities which suffer from basic needs problems and from the side-effects of modernisation, particularly pollution. We believe that our work is valuable in lending a helping hand to depressed rural communities so that they too can benefit from development. We are also able to give valuable feedback to policymakers and government authorities so that they can take actions to counter the ill-effects which are suffered by the rural population. We believe that non-governmental organisations like CAP can co-operate with governments to bring about genuine development at the rural grassroots level. In doing this, of course, we believe that genuine and equal participation by rural people themselves is very essential if they are to benefit from development.

Appropriate products

It is not only the public sector which should review its projects. The private sector is also guilty of a gross waste of resources and the promotion of lifestyles which lead to the alienation and cultural emptiness of modern humanity. Industries are producing more and more goods which are of little genuine usefulness and which may be unsafe, of poor quality and even morally objectionable. An example of such a product is the electric toothbrush, which I am told is getting more popular in the United States, where people may be too tired to perform the tedious hand motions required by the ordinary toothbrush. Needless to say, electric toothbrushes are luxuries which waste resources; we can only pray they don't enter Third World markets.

There are also products made these days to malfunction after a certain period, so that consumers will buy more of the goods. These products with 'built-in obsolescence' include everything from ballpoint pens to motorcars which rust and malfunction even when new. Our society is also swamped with books, films, comics and television programmes which overemphasise violence and sex.

In the choice of goods to produce, given our limited resources, we should emphasise appropriate products which first of all meet society's basic needs of food, health, housing, and education, as well as healthy recreational, spiritual and artistic pursuits. Products which we manufacture should be socially useful, durable, and safe to use, rather than being designed for attractive packaging and fashion.

To produce these appropriate products, we should not rely solely on modern capital-intensive technologies, but develop our own appropriate technologies which are small-scale, made from local resources, reflect the skills of our communities, and are ecologically sound. Indeed, there are already many appropriate technologies which have developed through the generations in our countries but they are neglected because of the faith our scientists and planners have in modern technology.

In the final analysis, however, nothing is going to change unless the people themselves realise the folly of chasing after all the modern gadgets of urban life. The advertisers have teased and tempted us to get more money to spend on more and more fashionable products—clothes, cars, video sets, electronic gadgets, and so on. Consumers have become the slaves of products. Products which are supposed to serve human needs have instead been used to create unhealthy artificial desires, and have twisted people morally and spiritually to yearn for gadgets and to slave their lives away to earn the money required to purchase them. So we have now to question our very minds, our very hearts, indeed our very souls: Is this what a human being is, is this what life is all about?

I, for one, do not think so. This is not the sort of creature which our Maker has designed us to become. A human being is a noble being, in possession of mental and spiritual faculties, who is made to live in harmony with nature, in friendship, love and cooperation with his other fellow human beings. We were not created to destroy nature and to compete with one another for fashionable goods.

Only if we change our attitudes, avoid the temptations of rapid modernisation and lead simple but happy lifestyles, can we avoid the wastage and great economic dislocations which result from the depletion of our resources. We must have a type of development which focuses on humans' material and spiritual needs. After our basic material needs are fulfilled through the provision of basic goods and services like food, housing and health, we should be free from the dictates of fashion and the consumer culture. This means more time for leisure, recreation, reading and spiritual development.

What I am proposing as an appropriate lifestyle is nothing extraordinary. It is just returning to a simple way of living, to a harmonious relationship of humans to nature, to our fellow humans, and to ourselves.

THE MYTH OF THE MODERN

Nsekuye Bizimana

✧

I was born in Rwanda, central Africa, in 1949. Before I went to Germany in 1970 to study veterinary medicine, I had thought it quite obvious that my country should imitate Europe in every respect. In our eyes Europeans led a life full of happiness and luxury.

I was fortunate to receive a scholarship to study abroad, and have now spent more than 25 years living in Europe. During this time I have lived as a student, industrial worker and scientist, both in the city and in the country. I have talked with many different people, young and old, rich and poor, employed and unemployed. I feel that through these experiences I have been able to understand the ways of the 'modern world'.

All this has convinced me that the main cause of the internal problems in Africa lies in the fact that we Africans still believe, even after 25 years of independence, that progress simply means copying white people, adopting all their methods, technologies and policies. Imitating the ways of the West will result in the failure of Africans to make their own discoveries on the basis of their means, abilities, history and traditional skills. Until the present we have seen the results of such a policy: not only have we failed to solve our old problems such as hunger and disease, but we have also added new ones, similar to those in the industrial world, such as loneliness, alienation and the destruction of our environment.

I believe that positive changes cannot occur in Africa until the African people lose their illusions about Europe. In writing about my experiences in Europe, I hope I am making a small contribution to dismantling these illusions.

Life in the West

After living in Europe for a long period of time, I became more aware of what Western society lacked, and the many aspects of Rwandan life that are valuable. Much to our surprise, we Rwandans felt very homesick in Germany. We had everything in Europe. We had considerably more money than at home, where we had practically none. We were well dressed, which was not the case in Rwanda. Each of us had not only a radio, but also a cassette recorder and record player. Not even loneliness made us want to go home, since we three good Rwandan friends had each other. But while we appeared to have so much, we did not have everything in Europe. We missed the sense of togetherness which was so much part of village life. We missed home. Home not only in the broader sense, but also in the narrow sense—the place where one was brought up and where one's personal history began. Although we had so many opportunities in Europe, we were still not quite happy. We yearned for the old black milieu and the simple, natural, uninhibited behaviour of Africans. We wanted to see what had happened to our village. Who had got married? Who had had children? A nagging voice inside us kept repeating: 'You've got everything, but you're still worth nothing in this foreign land.'

All in all, however, I had a pleasant time in Europe; the longer one is in Europe, the less one expects from people. As the Germans say, one develops a thick 'elephant skin'. Also, the longer I stayed the more evident the various aspects of societal decay became. One clear example is alcoholism. When we first came to Germany and noticed that so many Germans drank, we thought first of all that it was because they had a lot of money. However we began to see how alcohol abuse was in fact an escape from the stress and unhappiness so many people felt. One German claimed that 'although technical progress has made life easier, it has also created problems. Many people cannot cope with these problems and the reaction is to reach for the bottle.' I also noticed that Europeans have a more fervent inclination towards work than the Africans do. This strong work ethic, although considered most productive, has many negative aspects. Many people in Germany suffer from ill health because their bodies are under so much stress.

Loneliness

Loneliness, the main illness of 'civilised' people, is also connected with the problem of overwork. As Mother Theresa once said, 'The people of the Third World are derided because they can't feed themselves. The people of the industrial countries have a far worse kind of hunger, which they can't satisfy with their technological achievement: the hunger for love, security and community.' Thinking about these words, I find a similarity to what I saw in Berlin. This is clear in a comment of a German friend: 'You see these crowds of people—Berlin has two million inhabitants—can you imagine how painful it is to feel alone in such a crowd? One feels like one in two million—in other words, like nothing.'

Initially I did not understand my friend. It took me years and years to notice that loneliness is a problem of industrial countries and that many people cannot cope with it. And when one does become aware of the problem, one becomes aware at the same time that many people in these countries have been 'dead inside' for a long time, and that they are only 'moving body masses'. One starts to understand also why people take pep-pills, sedatives and alcohol, just to be able to talk to each other, and bear life. One must ask oneself whether or not technical development has brought these people more happiness or more sadness.

As a foreigner in these countries one first notices loneliness in oneself. Most of all one feels lonely when everyone is aware you are sad, but still nobody approaches you to ask what the matter is. One feels lonely when one realises that one has to cope with all one's problems alone. Conversations with many lonely people told me that Germans have everything except the most important thing: love and security. They are unhappy despite their material affluence.

Most people in Germany owned a dog. I saw the dog as a symbol of a brutalised industrial society. The dog is supposed to supply that which humans are no longer capable of, namely love, affection and security. Love for an animal turns into an exaggerated love of oneself and a hatred of other people.

I also observed crisis in the family. The concept of extended family has died out. The family has been reduced to just a few close members, and even these people have difficulties getting along with each other.

Some people are so tired of living that they commit suicide. Why do more people in rich countries commit suicide than in poor ones? Why are people unable to love each other in industrial countries? Why are there far more lonely people in these places? One keeps asking oneself these questions when confronted with this problem. The first reason, I believe, is the fact that the people are under so much pressure to achieve. If one works a lot, one has little time left over for other people, even if one is in need of company. A person who is perpetually under stress and who never gets the better of his work, cannot be at peace and have the friendly disposition which other people would like.

Technology has also invaded the home. People in Europe no longer entertain themselves. Television entertains them. In contrast, people in Africa do not have television and radio to amuse them and there is, therefore, far more contact between family members. Also village festivals lead to a strong feeling of togetherness, with everyone taking part, and without any commercial intentions. This feeling of togetherness is never apparent in large cities in Europe or America. In fact, in these cities people have very little trust for one another. There are reasons for this lack of trust. At the root of it all lies the way in which people from all walks of life have been thrown into a big melting pot. And with the continual moving from one flat to another, one remains anonymous and unnoticed, and thus people become indifferent to their neighbours.

Development and Breakdown in Rwanda

Applying the same economic criteria for judging development in Rwanda as in the West, the authorities in Kigali had reason to be pleased with the results attained up until the breakdown in 1994. In the rural areas the quality of housing had been improved from huts to stable houses, and primary hygiene practices had been instituted. Many industrial products (with negative effects that only became evident later) offered more comfort than the traditional way of life. More Rwandans were being educated at home and abroad, and the number of subjects taught at the national university had increased. In fact, Rwandans were proud in claiming that Kigali had become a little European island in Rwanda.

The fact is, however, that from the point of view of human development, a sharp deterioration in the quality of life had taken place. The influx of every imaginable Western industrial product, from washing machines, hi-fi sets and video players to sweets and cars, led to an even greater admiration of the West. People wanted to acquire these fine things more than ever, but only a small portion of the population could afford them. The result of all this was the creation of an arrogant bourgeoisie, who believed in the principle that 'the more you have, the better you are.' Such powerful bourgeoisies formed a state within a state, thus controlling almost everything. This was similar to the dictatorship-of-capital seen in the West. The benefits of modernisation were rarely distributed equally. Thus those areas with fine modern installations despised the areas without. As a result, tension developed between different areas within the country. This is known as regionalism. In many African countries, regionalism corresponds to tribalism, because different tribes live in different regions. Money reigns and thus people go to any measure to acquire it. This striving after money inflicted serious damage upon our culture. Egoism gained ground. Nobody did anything unless he personally intended to gain from it. Little was shared and the spirit of solidarity amongst our people collapsed.

Building on traditional ways

I believe that one way of resisting the forces of modernisation, and avoiding the consequent problems, is to decentralise political and economic power. If the new structures in Africa could only make it possible to return to the old form of democracy, one would sort out quarrels, and especially the most private of these, within the family instead of in court. Villagers would be able to assume responsibility for themselves and thus decisions could be made at a village level. Village gatherings would take place regularly, just as they did in earlier times.

The decision-making processes in African villages show that our ancestors could be democratic without belonging to political parties. The village chief did not simply issue instructions but first made his decision after a long discussion which normally took place with other village elders under a tree. Disputes between individuals were sorted out in *gacaca*, the Rwandan name for a village gathering, where everybody was free to give their opinion, and decisions and judgments were made in a democratic way. As they look towards the future, intellectuals in Africa should direct their thoughts towards finding a development model which would build on, rather than destroy, such traditional practices—which would, in other words, more closely correspond to the principles and attitudes of the African way of life.

A certain amount of cooperation with industrial countries is necessary, since their experiences can be of use to Africans in some areas of development. But development so far has done more damage than good. This damage is most evident in agriculture. The mechanisation and chemicalisation of agriculture has increased dependency on imports. Relying upon agricultural machines means that work comes to a halt if the machines and their spare parts can no longer be paid for. Employing machines in agriculture is also objectionable on the grounds that the rich, the only ones able to afford them, take away the land from the poor. The fact that precedence is given to the production of export-intended cash crops such as coffee, tea, cocoa, flowers—rather than to the crops which are needed to feed the indigenous population—soon increases the risk of famine. And the foreign exchange which is earned does not serve the masses but is used to acquire luxury products from abroad. I am of the opinion that one should first exhaust all agricultural techniques which are simple, cheap and harmless, before looking to Europe for ideas. A lot of imagination has to be applied not only in agriculture and animal breeding but also in techniques for preserving produce. For example, why not use ashes from banana leaves as a preservative for millet, instead of using DDT?

A clear concrete example shows the strengths of using local resources. In order to kill the ticks that carry East Coast fever, animals are driven into dips full of insecticides. Sometimes the animal dies from strong internal toxicity; at other times the insecticides penetrate the skin and are deposited in fatty tissues and the milk. This whole process would be unnecessary if animals remained in their natural localities, because there they are already resistant to such endemic illnesses. Furthermore, those owning only a few animals could remove the ticks by hand.

Africans have to be extremely careful before adopting new means and methods in agriculture. To the question of which organisational structure to apply in African agriculture, the answer is according to the African ideology: put simply, cooperatives should not be forced and large farms should not be formed. The existing agricultural system—that is, the traditional system—should simply be improved.

Moral signposts

Along with recognising the value of African ideology, each African country has to raise for itself a moral 'signpost'. It is important that Africans develop their own morality on the basis of their own traditions. The signpost should include positive traditional values (for example, respect for the elderly, fondness for children, a spirit of community, and so on), as well as condemnation of negative aspects of the traditional way of life (for example, superstition, monarchic tendencies, female circumcision). One should encourage Africans to find their orientation according to this signpost.

Religious ceremonies which are really a veiled means of obtaining and retaining power for some should be replaced by more traditional village festivals (celebrating births, deaths, harvests, and so on) which would allow the people to develop a common spirit and, in so doing, suffocate the encroaching Western ideology of 'each man for himself'.

Finally, I would like to direct a few words to industrial countries. They should not act as if the current events in Africa have nothing to do with them. There is only one Earth, and technological progress in the media and transport are making it smaller. Changes in one region have consequences elsewhere. It is difficult to imagine that, in this world of ever-decreasing size, certain regions continue to become richer and richer and others poorer and poorer. Policies based on egoism, whether on a national or an international level, are doomed to fail.

This paper is based on Dr. Bizimana's book White Paradise, Hell for Africa, *available from Edition Humana, Grainauer Strasse 13, 1000 Berlin 30, Germany.*

THE IMPACT OF MODERNISATION ON INDIGENOUS PEOPLES: The Case of Sarawak

Evelyne Hong

✧

M uch has been said and written about social change in the societies of the Third World. The conventional wisdom has been that colonial contact—and more recently development—would free the less fortunate in these societies from poverty, hunger and illiteracy. The development advocated was all too often equated with economic growth, a development concept which has proved to be a fallacy. In the process of imitating a Western model, Third World communities everywhere have been losing out.

The disintegration of traditional cultures follows a generally predictable pattern. Subsistence farmers with their own land and tools are transformed into wage labourers. Those that manage to hold on to their land switch to cash crop production, thereby becoming subject to the mercy of the market economy and the chain of traders and middlemen. The dispossessed, the unemployed and the young leave their indigenous communities for the bright lights and opportunities of the towns, only to become the victims of urban poverty and squalor. Traditionally self-reliant societies become dominated by outside forces that are more powerful economically and politically. Individuals and communities are left with little or no control over their existence and livelihood.

What follows is a case study of a Kenyah longhouse community located on the banks of the Baram River in Sarawak, East Malaysia. In recent years, these people have been confronted by powerful economic and social forces which have threatened their way of life and deprived them of their ancestral lands. This process began in the 1980s with the introduction of the market economy, formal education, and dam construction and logging operations in their forest lands.

Traditional Kenyah economy
The traditional economy had four important features:

- *Swidden agriculture*: In Kenyah society, swidden was once the main form of agriculture. It was based on a system of land rotation: land on which padi was grown for one season would be vacated and allowed to lie fallow while another plot was prepared for the next season's crop. There are a number of processes involved in swidden farming: slashing, felling, firing, sowing and harvesting. These major activities were done collectively by the community. Before these major agricultural activities were undertaken, the village as a whole had to discuss and agree on its commencement, as work on one swidden would affect neighbouring plots.
- *Land*: Since agriculture was the mainstay of the economy, land was the most important factor of production. Under the traditional system, there was no private ownership of land in the sense of aquisition through purchase. Under customary law, rights to land were based on felling virgin forest, occupying and cultivating the cleared land, and planting fruit trees on it. Rights over land which was uncultivated, on which no sign of cultivation existed, or which was abandoned and unclaimed, lapsed to the community. Usage rights in land could also change hands temporarily. The 'lending' of land (or usufruct rights) was a common feature among community members.
- *Labour relations*: There were two main forms of labour organisation in traditional Kenyah society: community labour and *adet senguyun* (mutual help). Community labour was used in building the longhouse, and on social and festive occasions such as funerals and weddings. These activities would be concluded with a meal in which everyone was obliged to

partake. Activities of this kind expressed community solidarity, and were sanctioned by *adet*, a traditional philosophy covering economic, political, and social aspects of Kenyah life. *Adet senguyun* was a system of reciprocal labour exchange among groups of families in the longhouse. It was employed for work in the swidden, in building a family room or a boat, or for any other activity which required help. It was an occasion for companionship and chatter, and was one of the main factors enhancing village cooperation.

There were also other forms of labour in swidden agriculture, namely *corvee* or obligatory labour. *Corvee* labour was performed by freemen for families of the aristocracy. This service was provided during the main stages of the swidden cycle.

* *Distribution of surplus*: Due to the low level of technology and the perishable nature of food, a food surplus was not easily stored. It was instead distributed through a form of generalised reciprocity or gift exchange. The more an individual gave, the more he was appreciated and respected for his generosity. This element of reciprocity in consumption enhanced the spirit of sharing sanctioned by *adet*. The aristocracy also observed this principle of reciprocity with their freemen. If a freeman's crops were destroyed, if he experienced a bad harvest, or if he could not complete his room, he could be assured of assistance from the aristocrat. Kenyah moral order was based on such relationships of reciprocity and interdependence. This system succeeded because credibility and purpose were ensured by the *adet* system. The gift exchange between freemen was also a leveling mechanism by which excess food and labour services were shared. This helped to minimise economic disparities between freemen in the society.

Introduction of the market economy

Contact with the market economy, and the legal, political, social and cultural institutions related to it, significantly transformed the Kenyah system of production. It was the introduction of rubber cultivation that initially linked the communities along the Baram River to the fortunes of international trade. Once planted, rubber trees became a permanent feature of the land, and longhouses no longer shifted with the exhaustion of swidden farms. Now, the Kenyah people had to stay close to their rubber plots to look after them, and could not move as freely as before.

Cash crop cultivation also led to a more individualised system of production. Less cooperative labour was required since rubber, pepper and coffee could be worked efficiently on an individual basis. And unlike the major swidden activities, work on one farm did not have an economic effect on neighbouring farms. The result was less mutual dependence between families in production, less collective decision-making, and less labour exchange. Increasingly, each family concentrated on its own cash crop production and kept the fruits of its labour for itself.

In 1956, the state initiated subsidies for cash cropping. These schemes were introduced to discourage swidden cultivation and to encourage the development of a more sedentary rural population. These efforts drew the community further into the larger economy and society. Concentration on cash cropping also meant less time and labour expended on growing food for subsistence. When there was not enough, food had to be purchased.

Some of the most profound impacts resulted from changes in the land system. The classification of all virgin forest land as State Land in 1949 checked the further expansion of traditional swidden holdings over which families could gain prompt rights of use. As a result, less swidden could be grown, or the period of fallow had to be reduced. Land problems were made more acute by the fact that the best lands were frequently planted with rubber.

In the 1960s, State Land was leased out to timber companies. Logging camps and sawmills mushroomed all over Baram as the timber industry boomed. Employment was now available to the Kenyah, opening the possibility of earning a wage as an alternative to farm work. Many young Kenyah moved out of the traditional economy and its set of customary laws.

With increasing population, there was a need for more land; but by now, it had been made scarce by land legislation and the exploitation of timber. Rubber had to be sold, and the further the rubber lands were from the trading *pasar* (bazaar), the less advantageous it was for the owners. Due to these factors, many communities began to abandon their homes and swiddens upriver to be nearer the *pasar*, the hospital, and the school for their children. As land downriver was prone to flooding, there was a shift from swidden agriculture and self-subsistence to swamp rice cultivation, which often had to be supplemented with purchased rice for the family's needs.

The shift to cash cropping from padi cultivation and the increasing monetisation of the economy had a great impact on social relationships within the community. The aristocrats who owned more land were able to engage successfully in rubber cultivation. Income from rubber was reinvested into more rubber land, and wage labour was employed both within and outside the village. Tribute labour lost its significance because hired labour also performed the job. Wealth was measured in terms of rubber land. The aristocrats could afford to engage wage labourers, and in some cases discontinued swidden civilisation on a subsistence basis. Cultural practices (such as *adet*) associated with swidden were abandoned by these groups.

Customary rights still held on swidden land, but not on rubber land. Rubber land became private property, with the owners holding permanent rights of ownership, cultivation and disposal. Owners could pay hired labour to cultivate the land, a system of wage labour which had never been part of the traditional system.

The use of surplus

With the exposure of the village to the market and the increasing commersialisation of the economy, the method of utilising surplus changed. The aristocrats could exchange surplus rice, and later rubber, for a large range of consumer goods bought from the *pasar*. Surpluses could now be converted into money to buy furniture, refrigerators, town clothes and bicycles, most of which were prestige items and manifestations of conspicuous consumption. Money could also be used in gambling. People could travel to the *pasar* to buy goods and tinned foods, collect rent from their properties in the *pasar*, go to the cinema, eat in restaurants or entertain guests. Cash could now be saved in fixed deposits in the banks. Thus, more money meant more consumer goods and a 'fancier' style of life.

The village economy was no longer self-subsistent. Outlets for surplus existed in luxury consumption or profit ventures. This commercial attitude of the aristocrats led to a breakdown of reciprocal relationships. Freemen could no longer turn to them for rice or help, as every dollar could now be used for investment and profit.

In one revealing incident in the village, a sick man had to be sent to the hospital downriver. Instead of bringing him to the hospital on his own boat with outboard motor, the village chief suggested the man be sent on the slower commercial launch. This would not have happened in the old days.

Mercenary considerations have taken over in other ways, reducing community spirit in the village. A few of the better-off farmers have stopped participating in work parties and are now hiring workers to perform tasks in the swidden and rubber plots. These farmers, now involved in business, have opted out of the mutual help system of *adet senguyun*. Meanwhile, farms have become choosey about selecting help for the work parties. The sickly, weak or old are less than welcome because they can contribute less.

Money relations have also replaced customary social ties between the aristocrats and the freemen in the village. As a result, the distribution of wealth and income has become more unequal. Aristocrats have been able to enjoy higher levels of consumption at the expense of this inequality. Reciprocity between freemen has also broken down. Surplus meat and fish can now be sold to the timber camps instead of being distributed among the families.

Culture and lifestyle changes
In traditional Kenyah society, the longhouse community formed the total world in the sense that this was where all the spiritual,

cultural, and physical needs of its members were provided for. The importance of the longhouse was further strengthened by the wealth of rituals and ceremonies connected with the agricultural needs of the Kenyah. It was through the participation in these activities and rituals that Kenyah socialisation took place. All this had an ideological base in the *adet* system which maintained community values and the rules of customary law.

As the traditional self-sufficient economy unraveled, the Kenyah people became transformed into members of a consumer society. A taste for new products and services was acquired. Sugar, cigarettes, beer, clothes, transistor radios, TV sets and refrigerators became indispensable. Mothers gave up breast-feeding and began bottlefeeding their young. A new interest in accumulating money developed, as the means of acquiring goods, services and property from the outside. The 'good life' could now be had if one established links with wider society. For those who did not make it themselves, the dream was transferred to their children. So they worked and sacrificed to send their young to school.

Education

With compulsory formal education, the most important function of the community is taken over by the state. Kenyah socialisation is replaced by the schooling process, in which the child is systematically taught not only reading and writing skills but patterns of behaviour and values appropriate for a society based on competition, achievement and reward. For those who succeed in the primary school, secondary education is extended into boarding school in towns, and the child is given a taste of urban life and comforts. The youth who passes through this system is cut off from his traditional community. Education thus draws the young into the modern sector, initially through the school system and later through the labour market. The training and discipline a Kenyah child undergoes in school serves to alienate him from traditional community life.

Those who succeed in school outside the village are a great asset to the family. For those who fail, all the years in school are wasted, and the sacrifices made by the family are bitter experiences. Parents have not only lost the gamble, they have also lost their children. School has not taught them to be farmers, so they cannot help in the swidden or even the rubber farms. They are thus neither qualified enough for a white collar job nor equipped for agricultural work. They are ashamed to return to the villages where they would be deemed failures. Unable to fit into the life of the community, they are more comfortable in the *pasar* doing odd jobs or working in the timber camps.

All over the Baram, longhouse communities are devoid of youths, who have flocked to the timber camps and the towns. Those who come back to the longhouse during holidays find it a dull place. They miss the cinema and the shops in the *pasar*. The conditions for the breakdown of the traditional Kenyah longhouse community in the future have been created as the traditional system loses the ability to reproduce itself.

To the successful educated Kenyah, the town is where the future lies, and they are ready to cut lines with the traditional community. It is very common to hear Kenyah girls saying that they do not want to marry a farmer. Marrying an educated Kenyah male means freedom to leave the longhouse and the traditional ways. Educated Kenyah males want wives who can speak English, dress well, walk in high heels, and drive a car. Traditional habits are discarded for the modern ones they share with their urban counterparts. Education thus makes the process of Kenyah deculturisation complete.

Although this study is based on a particular ethnic group in Sarawak, it is significant to note that the traditional communities in which these changes are taking place are relatively remote. Yet cultural change has come to stay. Given this fact, the degree of change occurring in areas which are not as inaccessible—and are thus more exposed to the market economy—must be even more profound.

Waking up from the development dream
The above analysis has shown how market forces and the ethos that accompanies these forces have transformed the traditional

economy and cultural values of the Kenyah. We can see similar conditions in many other traditional communities, where agriculture is being abandoned because there is no one to work the land, where those who can would rather join the market labour force working for wages. The replacement of subsistence production with cash cropping has made possible a form of consumption based on the market economy and the use of cash, in which more money allows for more consumption. This contact with the market economy has led to the breakdown of value systems in traditional societies.

The market economy generates a pattern of consumption which is both compulsive and necessary for the economy to sustain itself. In this context, a whole propaganda machine is needed to make one consume more. Whether this consumption is really necessary or irrational does not matter. This is the consumer culture in which 'the market has to be moulded to suit the product'. Ultimately, the 'homogenising of consumer tastes' results in the creation of a universal culture which is everywhere the same—in food, dress, song, leisure, housing, etc. Universal culture also means that since everywhere the same types of goods and consumption are being sought after, producers can enjoy economies of scale, as their products are identical the world over.

This universal culture is Western-based, as the products it promotes and the lifestyle it encourages are Western and 'modern'. This brand of consumption and lifestyle furthers the interests of the transnational corporations whose products are thereby promoted. As every city, town and village undergoes the process of becoming modern, they begin to share the same colourless anonymity that modern cities share the world over. The four-lane highways, skyscrapers, shopping complexes, condominiums, cinemas, supermarkets, discos—this is the cosmopolitan city, devoid of any indigenous character, culture or charm, and a very lonely place to live.

Advertising propels this culture even further. In most Third World countries (including Malaysia), the largest advertising agencies are American. In the advertisements that adorn the glossy magazines and the television screen, the man who smokes a particular brand of cigarettes is always successful in life and with women. The girl who uses that particular perfume is always admired and loved. This TV set is used by the most popular film actress. Images of luxury are used: swimming pools, golf courses, sleek cars, airplanes, beautiful sophisticated women. The impression is that a little of the glamour and success somehow rubs off when the product is used.

Even more subtle is the other message put across all too often, that indigenous ways are to be discarded, even held up to scorn and ridicule. Indeed, throughout the Third World, traditional culture has become a negative reference group, a group that all ambitious go-ahead people seek to escape and deny all connection with. This is especially so with the younger school-going generation who have become totally displaced and alienated misfits in their own society. The school system after all trains them to fit into the efficient slots that modern society has been built upon, be it as clerks, engineers or accountants. There is no other definition for success. One is 'modern'—in dress, tastes, behaviour, employment, lifestyle—or one is a failure.

One of the most effective and successful purveyors of this dominant culture is tourism. Tourism clinches the images of the good life the locals have only heard about or seen in celluloid. The tourist's money and his lifestyle and the amenities that are created to cater to his 'needs' are as large as life. Tourism has an aura of its own: big beach hotels, swimming pools and casinos, skiing, yachting and unquenchable fun. Associating with it—as waiters, bellhops, cashiers, laundry men—also pays.

The kinds of aspirations resulting from Western cultural domination cuts people off from their cultural moorings, alienates the younger generation from traditional indigenous society and destroys people's self confidence. Cultural domination distorts the personality and engenders a sense of inferiority.

This sense of smallness and inadequacy can be extended to nations as well. When nations do not believe in their indigenous ability, knowledge and ingenuity, all that is traditional is thrown away (except when it is for the exotic approval of the tourist).

We replace it with all that is Western and modern. We mimic their dancing, music and art forms, while traditional theatre, song and dance is discarded with shame. We copy their cities, traffic systems, buildings, science and technology, health systems and education. We have simply lost faith and confidence in our own abilities to create and build for our own needs. What is worse, we perform this wholesale imitation in the name of, and at the expense of, our people and our communities.

Although it would be simply pure nostalgia to hark back to some concept of a romantic past, there was much good in the traditional society: the community spirit, the sense of sharing, the reverence towards the land which was the source of livelihood. In talking of 'development' then, these are the cultural values we should work towards, rather than those of the modern consumer culture which is both alien and alienating.

Many people in the West have become disillusioned with the soulless individualism of their system and are looking towards more community-based ways of living. It is ironic that they are often rediscovering it in the countries and cultures of the Third World, while we in the Third World are in fact rejecting and destroying community values and reverence for nature, replacing them with the the worst there is in Western society. At its best this aping of the Western model remains a caricature of the original. At its worst it destroys the minds, souls and even the physical bodies and lives of our people, as is evident in the existence of 400,000 drug addicts in Malaysia.

Development and production in the Third World should be one which is meaningful and self-reliant. It should harness the use of appropriate technology and conserve the environment. It should also inculcate the values of community spirit, humaneness and sharing. Despite the onslaught of Western systems of production, habitat and culture, there is still much in our society that we can save and build on. In the villages, there are still some common people who hold fast to their community values, their harmonious understanding, respect and love for the natural environment, who practise the appropriate technology and lifestyle which others talk about, and who are unhappy with developments in the modern world. Hitherto these common people have been looked down upon as having conservative values, as being obstacles to development.

We ourselves must wake up from the development myth and dream, face the fact that modernisation is alienating our human values and destroying our resources. The deepening ecological crisis that threatens our very existence has its roots in the economic model upon which contemporary western civilisation is predicated, and requires us to question the very basis of industrial society. Therefore, we must rediscover our own identity, and re-build our societies along our own lines, based on community spirit, respect for nature, appropriate forms of habitat and technologies, simple lifestyles and finally—most important of all—based on human, really human values.

IV

Society and Ecology

INSIDE NATURE

Sigmund Kvaløy

✧

The debate in Norway over whether we should join the European Community has taken so much time, effort and emotion that it has been like a mental a block, a constraint against writing for this collection, even though I have very much wanted to do it. Many of the participants at the 'Future of Progress' conference are striving for the same thing that I am, and have the same analysis and passion.

Reflecting on this, my thoughts wander to A. N. Whitehead, whose ontology takes facts as emotionally imbued and value-laden. This is a radical departure from the Western mythology of a quality-free reality, and is an integral part of his process philosophy.

The task of philosophy is to look through the grid of conventional concepts. The task of eco-philosophy is to do so in a way that strengthens the observer's roots in the Earth. The grid to be penetrated now is one that has grown dangerously abstract in relation to human needs: the urgent task now is to regain *concreteness*. To Whitehead—as well as Buddhists and Western process philosophers central to the development of eco-philosophy— the concrete world is a value-saturated, creative process.

The value-laden array of facts spread before me today, like any day, consists of several layers. One of them contains the elements for my discursive writing, another has to do with the priorities of my daily existence, a third one concerns my life in total. The second breaks into the first at short intervals, and once in a while the third erupts.

We normally keep these layers strictly separated, and our academic tradition trains us to do that. Process philosophy teaches us to be at least *aware* of how these layers function and mix, and how we can benefit from that awareness. When we are considering a crisis of existence, it becomes a necessity to let the layers of our life's value-laden facts interpenetrate, because what is needed is a *total* grasp.

This juggling of the layers of my life-stream has now pushed the political struggles over the European Community to the background to make way for reflections on 'sustainable development'—a star-studded catchword ever since the publishing of the Brundtland Report. The concept is so feebly non-defined that it has led politicians and the public straight back onto their mechanistic track. Brundtland only talks of material conditions. But concrete experience all over the world should have taught us that sustainability has to start with human society and *its* sustainability, because it has become abundantly clear that the constitution of that society determines how nature and her gifts will fare. And I want here to stress as strongly as I can that the few European process philosophers that we know of—Hegel, Bergson, Whitehead, Sartre, Merleau-Ponty, our Norwegian Dag Osterberg, and a meagre additional few—ought to be compulsory reading for the people who yearn for a radical and concrete alternative to the cliché-like abstractions of Brundtland.

Processes of sustainability have to activate all the layers of human existence as part of nature's existence, otherwise these layers will in due time mix abruptly, leaving us with shreds in the wind and chaos instead of sustained creation. Experimenting, in this vein, at being myself and simultaneously observing myself, I shift right now from my overloaded EC situation to the eco-philosophical reflections asked of me by this book.

Inside and outside

Right now, I am typically torn between two alternative action programmes for my day, the first day of 1992—the last year before Brussels turns loose the inner market of the EEC. I can either strap on my skis and wander off to a little cottage in the Saetereng forest, or stay here with my typewriter. The first tempts me strongly, since it quickly leads me into Nature. As I know from old experience, that course would enable me to enter a realm of intense happiness—the concrete, inside 'life world' of the human individual. The feeling springs from a view opposite to conventional thinking, where we go *out* into nature. My world-view is one in which we move *inside* when we leave the modern concrete house or the city (or the EEC) and enter Nature.

The alternative to that is to stay 'outside,' without my skis, in the dry and cold realm where the typewriter belongs, communicating with other, similar parts of this outside world. This is the sort of communication that presupposes the irrelevance of emotion and value, by accepting a world that is an abstraction in relation to the concrete life-world of human beings—the world that made them and keeps them living. This is the thinking of Whitehead and his initial inspiration, Bergson, and in general of all the process philosophers. But how is this relevant to a conference on ecologically oriented development?

The Euro-Cartesian worldview

The human being grows and expands and matures through movement of her body, through concrete activity. The wonder of the human being is not her intellect, but her endlessly complex body (or 'body-mind', if you must). Without the body and its movements in concretely moving nature, the world would have no colour or sound or touch or shape or rhythm. Without qualities like these, no sense of beauty-versus-ugliness could be built. Worse, there could be no sense of good-contrasting-bad, and worse still, no quest for comprehensive meaning. In other words, without the human body, there is no aesthetics, no morality, no religion, and no philosophy. This is why 'artificial intelligence' is the wrong track: the computer can have no body. Altogether, this concept of the essential, organically created body is the extreme opposite of the mainstream, Cartesian, Euro-American (and more and more, Japanese) worldview.

The Euro-Cartesian culture is a unique phenomenon in human history, in that its world is static. That fact is reflected right now here at the Saetereng farm, where some carpenters are erecting an enclosed porch to my 200-year-old house; I struggle with them every day, because their straight angles contrast so sharply with the soft, rhythmic lines of the old building. They are reflecting the influence of the Western, outside world, while my house speaks to us of being inside Nature. 'Inside' is, in fact, where the global eco-social crisis is now forcing us to return, so I say to my carpenters: 'You are already old-fashioned.'

Luckily, they understand, since they and I live in a community which is full of buildings attuned to Nature. Carpenters in Oslo—the Norwegian bridgehead of what I call the 'Advanced Competitive Industrial Dominion,' or ACID—would not understand. But the major part of my country-folk do, and that's also why we still have a majority which is against moving 'outside' to be lost in the static desert world of the so-called European Community.

The northwestern part of my house, built from heavy timber, has spent two hundred years sinking slowly into the ground, and doing it with grace. It was built rhythmically, without architects. The carpenters—torn between two worlds—say they want to jack it up and make it straight. Without knowing, it they want to stop time, and make the house conform to a Platonic ideal. But since they are not completely caught up in the European-Cartesian world, they understand when I say that a house should wither like everything else in Nature.

Then we discuss the EEC, and they express hatred towards our prime minister, Gro Harlem Brundtland, who is the main individual pushing Norway 'outside' into homelessness, to become an integral part of the European Superunion. She is the head of the World Commission for Environment and Development, too, and she believes in economic growth to cure our Earth's illnesses. She is one of those barriers trying to stop Time, which is also Nature. The carpenters hate her, because her programme would eliminate their existence as small mountain farmers.

It encourages me that it still doesn't take much coffee and pipe smoking to make these connections clear in my country. The carpenters and I reach an agreement that the northern part shall be permitted to continue its sinking and that we shall try to hang on a few more historical seconds, surviving as the EC pyramid crumbles. They have left, it's snowing softly, it's night and I am listening to Billie Holiday. She is the future, sinking into the earth. Inside.

The parts of my own Norwegian nation-landscape that have preserved the inside relationship to Nature are essentially parts of 'the Third World'. Twenty years ago I arrived in another part of the Third World—the Plow Furrow Valley of Nepal— where within a few days I was drawn into the untourist-like task of getting a female yak out of the scree. The contact with their world was immediate and compelling. Many years later, a youngster from the valley spent a half year at my Norwegian farm, and then some time in Oslo. He concluded: 'In all the things that matter to people and their animals there is a closer relationship between your village and mine than between your village and Oslo.'

Meaningful work

Oslo, our capital, is ACID's bridgehead in Norway, and shares ACID's uniqueness among all humanity's cultures. Whitehead, Bergson, and a few other Western philosophers that I mentioned earlier tried to break out of that uniqueness, and in the attempt they felt compelled to develop a language that is almost incomprehensible in Western terms. The languages of ACID presuppose a static, quality-free world, where individuals grab and dislocate objects.

The basic difference has to do with meaningful work as the very basis that society is built upon. The Third World has that, the First World has lost it. Meaningful work is a constant interchange with living nature. Thus it has to be a creative process— with human beings inventively meeting challenges to the totality of their bodies, maturing throughout their lives into the manifoldness of Nature. Since Nature's body is an enlargement of our own body, just separated by 'semipermeable membranes,' this kind of maturing brings deep satisfaction.

Maturing of this sort can only happen through serious work— through activity that is necessary for your material survival. In that respect it is basically different from leisure activity or participation in Western-style education. Human maturing has to happen as a response to compulsion from a non-human authority with which you cannot argue. In a viable human society, Nature sets the ground rules while politics—human rules—are secondary.

In the First World, it's the other way around, due to our over-powering technology. Material abundance has the effect of putting politics first, and human maturing does not occur. In other words, a genuinely human society will blossom only if its resources in materials and energy are meagre and hard to extract, and where the methods available to extract them entail the complex use of the human body in direct and concretely active involvement.

All of this means process as reality, and the mind-body dichotomy as a misunderstanding. And the Western dictum that 'ethics comes first' is also a misunderstanding: our values we have in common, what divides us is our worldview. The latter is what needs to be cleared up, and then morality falls into place by itself.

Let me now give a more formal definition of 'meaningful work':

- It is an activity necessary for the person's material existence, giving it a direction and practical seriousness not shared by any other human activity.
- Its products (material objects, services and various thought structures) are such that they do not cause damage to the continuance of life's organisational complexity (either in the ecosystem or in human culture) without time limits.
- It poses enough challenges that the potential talents and capabilities in the human individual and her/his group are brought to bloom.
- It demands of its partakers the building of solidarity and loyalty, as well as the practical techniques of cooperation.
- In general, it engages children, not as play only, but in ways needed by society.

Education through engagement

This last point is important, since I am now giving the social-ising and naturalising role back to work, removing it from the outside realm of schools and Disneyland leisure where ACID— the great aberration—has put it in the last generation. The main elements constituting human personality are put in place before the child reaches the age of six or seven. If personality is mainly built through meaningful work, then children have to partici-pate in work.

Here my opponents will object by referring to the classic example of children pulling carts through the British coal mines—a picture that was in everyone's history book. I counter by describing child-rearing among the Sherpas and among my grandparents in Norway. In those cases, the tasks given to chil-dren were always quite carefully selected to ensure that the chil-dren would succeed and thereby build up self-assurance. These child-rearers were not part of ACIDic society, so it would never occur to them to put a child in front of coal cart. Without knowing it, my critics are opposed to ACID, not to tasks that earnestly treat a child as a responsible and highly valued member of society.

My contention is that we have removed our children from a trusted role as serious contributors to society's survival, treating them instead as playthings or investments in a remote future through training programmes outside of society and nature. This goes a long way toward explaining why we are left with a society that is sliding unawares into eco-social catastrophe. Steering away from this path will require people with identity within their own culture, self-assuredness, inventiveness in the practical sphere, originality of approach, and willpower. We have systematically removed the training basis for such individuals, who—anchored in their culture—function well only in socially cooperative contexts. What we have instead are masses that are easily molded by commercial mass media—for instance to say 'yes' to the EC just because of its cheaper sirloin and the possibility of getting cheap liquor everywhere.

ACID is a type of society where work (or 'employment' in

this case) does not fulfill the above-mentioned characteristics of meaningful work. This is very revealing of ACID's nature, since it puts into sharp focus how it is an aberration: by and large, human societies around the world have had their economy based on meaningful work.

ACID removes its citizens from a concrete relationship with nature and teaches them to accept being clients of an abstract dream-world. Contrasted with that, societies based on meaningful work teach their members in a systematic way from early childhood to understand and feel their manifold dependence on the natural world, and so teaches them to take care not to harm Nature.

Another aspect to take note of regarding such a society is that it flourishes on meagre and hard-to-get-at energy and material resources. An easy abundance of such resources tends to put it on the track towards its own destruction. This is a course that brings society away from an inside relationship to nature.

A glimpse of life inside Nature

On a simpler level, I have just experienced part of such transition—from the 'inside' of nature to the 'outside' of my typewriter, having returned on skis from work in the Saetereng forest. I have been cutting birch trees for firewood and at the same time thinning the forest so that the little trees may receive more sunlight and space for their roots. The snow here in this part of the world is still full of animals and their tracks, and today I saw a mother moose with her calf. She had used my frozen ski tracks from yesterday to reach a better feeding place and to protect her child from struggling in the deep snow.

Where I had felled trees yesterday, the snow was marked by the mystical but purposeful patterns of a family of hares that had been feeding on the twigs of the trees I had made available to them. Inside that pattern, there was also the very purpose-laden tracks of a red fox, so I must be making things available for his sustenance, too.

Nothing around me is the same as last year, not even the same as yesterday. The snow bends down my trees: the mountain ashes, the alder trees, aspen, willows and their smaller relatives, the withie—I even have a few clumps of bird-cherries. They all bend differently. They will be permanently shaped by it, and while I am skiing past them, I try to guess how they will appear when spring has arrived, and how they will grow into a new balance.

I just said I *have* these trees, but I don't—it's more like they have me. I know all their places and their particular histories of the last fifty years, and that has given me an idea of what they are trying to do and what relation they bear to the various families of animals, birds, and insect colonies that together make up a place on this earth that has got me. It's a place that I can never leave, because most of the activity that is my activity, my body's involvement, is actively part of the process that is this forest even when I (i.e. another body-activity of 'mine') am removed to the Plow Furrow in Nepal. When that part-person is way, the trees and the animals and the flowers I have planted around the farmhouse are all taking care of my activity-stream here. And it's also taken care of by my human neighbours, busy with projects that are also my ongoing projects.

The skis I am using are the wide jumping skis that were once made for me by my uncle, the one who took care of Saetereng before me. I remember trying to help him, but he never gave me any specific instruction. If I asked for guidance, he just told me to keep my eyes open and to try to develop an awareness and alertness that ACID's children—through their dependence on instructions or on machines to do things for them—never get. These days, I use my wide 'uncle skis', because they don't sink deeply into the snow. I need that since I have no fixed tracks, but have to go to different places to do my work every day. Sliding home in the evening, I surf on the top of the snow! Down the hillsides, curving around trees, I am in for new challenges and surprises every evening.

I am back now in the outside world, struggling at the type-writer, searching for words that I often don't find. My own language, my mother's *Trønder* dialect, is a language of the meaningful forest work that I am trying to describe. I cannot do anything more than try to give my readers a glimpse or hint of life in a particular, concrete society that even today possesses the much coveted quality of sustainability.

Humans as machines: a standardised monoculture

The 'basic need' that is the key term of the Brundtland definition of 'sustainable development' is never defined in the Brundtland Report. But we may approach what the authors had in mind by leafing through the book and noting every place they talk of what is needed. All they refer to are quantities of materials and energy and the mechanics of keeping up people's health. They might as well have talked about machines: the latter also need fuel, lubricants, shelter and technicians to repair them when they fail. There is nothing in the report that connects sustainability to anything specifically human.

Judging from masses of human experience reaped from many cultures, economic sustainability is dependent upon the existence of local, living cultures that have an inside relationship to nature. I would also say these are cultures that are based on meaningful work.

Let me put it this way: a basic, specifically human need is to grow and mature into a specific identity. That again means the living availability of a particular cultural tradition—a pattern of activity, thought and feeling that has emerged over many generations as the complex answer to the specific challenges nature offers in this culture's place on the earth. These local challenges are the main reason why we have a variety of deeply different cultures around the world.

A human identity worth the name presupposes an existence that extends in an unbroken manner far beyond an individual lifespan. Identity for the individual means a history and an inherited worldview containing everything that is necessary to deal with 'survival-with-abundance' in her own place. The catchwords here are *self-sufficiency* and *self-reliance*. Without identities reflecting the eco-social histories of the many different landscapes that Gaia presents to human beings, there would be no individuals or societies with sufficient moral strength and inspiration to counter difficulties as they arise. Neither would there be alternative ways of doing things. Alternatives are at the base of fruitful dialogue and discussion inside of mankind.

But these basic survival qualities are exactly those that the Brundtland approach threatens. That approach, in line with systems like GATT and the EC, presupposes a standardised human monoculture. Its mechanistic world model and its view of human individuals as identical elementary particles express a misunderstanding of both Gaia and her human children. The accelerating global eco-social crisis that we are now experiencing is, in sum, a devastating proof of that misunderstanding. One particular philosophical tradition, that of the Platonic-Cartesian brand, is now in for something unprecedented in the history of ideas: it is being decisively proven wrong by Gaia herself.

It is a fantastic irony that Gro Harlem Brundtland is named by the Western media the 'world mother of the environment.' If her definition of sustainable development had taken into account what are specifically human needs—the needs of having an existence as Gaia's insiders—the Brundtland Report would have to be entirely rewritten. To top it all, Mrs. Brundtland is now using her green media reputation to convince the crisis-frightened Scandinavians that they should leave their inside identities and move outside to become elementary particles of the European super-pyramid.

Almost miraculously, the Norwegian ecopolitical community has sustained itself despite thirty years of massive 'Brundtlandising'. It must be one of those marvels—beyond mechanistic thinking to grasp—that characterise Gaia's inventiveness. But Norway is little, so to all readers of these pages: we need your support!

LIBERATION ECOLOGY

Nicholas Hildyard

✦

The chief characteristic of the modern market economy is that it is exploitative. It is exploitative of the environment and it is exploitative of people. Its exploitativeness of the environment is now so well-documented that there is no need to restate the case. But its exploitativeness of people has all too often been set to one side as a separate issue.

It is time for this issue to be moved to centre stage, and for the environmental movement to begin to make common cause with social activists, both in the South and in the North. It is time, in effect, to recognise that while the environmental movement can usefully continue to bring environmental problems to public attention, it cannot hope to address them without also addressing the more fundamental issue of social justice. And that will mean confronting basic issues of how society is organised, who controls resources and decisionmaking, who benefits from current social and economic policies and who suffers from them; and how those policies can best be challenged.

Rubbish people

In the South, the intimate connection between the ecological crisis and the broader issues of social and economic justice is clear. Thousands are daily thrown onto the human scrap heap in the pursuit of short-term economic and political gain. Their forests are logged by companies whose only interest is to make as big a financial killing as possible off a single cut; their rivers dammed or polluted to promote industrial development which does not benefit them; their lands taken over for growing crops which are then exported to the North; their daily lives burdened by debt, insolvency, and often, political oppression. One New Guinea villager, threatened with resettlement through a dam project, poignantly expresses the sense of oppression that many now feel in the South: 'We have become rubbish people.'

Here, the problems of social justice and environmental destruction are demonstrably linked to the imposition of development policies that are undermining the livelihoods of the many in the interests of the few. In the North, the link is not as clear, but is no less intimate. High material standards of living— a level of prosperity that is only made possible through the exploitation of the Third World—cocoon the majority of us from the destructiveness of modern industrial society. But, as in the South, that cocoon is not available to the poor. However passionately those living on the dole in a high-rise estate may care about environmental issues, there is often little they can do about it. Their poverty denies them the 'consumer choice' available to the better off; the 'Perrier' option is not open to them. Similarly, the threat—or more often the reality—of unemployment means that the poor often have little choice but to accept any available employment and are frequently wary of demanding healthier working conditions for fear of losing their jobs. Indeed, the poorer the community, the more it is viewed as a dumping ground for polluting industries.

Recently, the California Waste Management Board paid a Los Angeles consulting firm, Cerrell Associates, some $500,000 to identify those communities that would be least likely to resist 'Locally Undesirable Land Use'—the industry's euphemism for toxic waste dumps. Such communities, according to the study, are characteristically rural, poor, politically conservative, 'open to promises of economic benefits', poorly educated and already involved in 'natural exploitative occupations', such as farming, ranching, or mining. In effect, waste companies would be best advised to 'target' communities which in the consultant's view, are too stupid, too disorganised, too poor and too respectful of authority to resist the siting of dumps which would not be accepted in richer, better educated, 'professional' communities.

Clearly, the North also has its rubbish people.

The Cerrell study is revealing for more than the insight it provides into the cynicism of modern corporations. Effectively, the question that the waste industry demanded of Cerrell was 'What makes a community vulnerable to exploitation?' Cerrell could have answered 'low income' and left it at that, but it would have been an incomplete and misleading answer, for there are many low-income communities in the United States which have doggedly resisted the siting of waste dumps.

The study's sophistication lies in having correctly identified low income as just one of a cluster of factors that collectively disempower a community—that so marginalise people economically, socially and politically that they see no choice but to participate in a system that is exploitative not only of the environment but of their health, their children's health, their community and their self-esteem. It is that state of disempowerment, of marginalisation, that sets Cerrell's 'target' communities apart from others; and it is a state brought about by forces that cannot be reduced to simple material deprivation.

The process of disempowerment

Disempowerment is not a recent phenomenon. In the North, the process has been inextricably linked with the emergence of the market economy; the replacement of the informal economy by the formal economy; the introduction of wage labour; the alienation from communal use (though not necessarily communal ownership) of land and other key resources; the undermining of local political structures through the encroachment of the state; the adoption of centralising and resource-intensive technologies; and economies of scale.

Disempowerment is about more than being denied the vote or nominally democratic structures of government. For real power—the power that gives a community a significant role in decisions that affect its daily life and its future—involves more than simply putting a cross on a piece of paper every five years. It demands that the community as a whole be able to exercise meaningful control over its economic, social, and political affairs.

Disempowerment results when that control is effectively taken from the community and vested instead in people or organisations which are either unrepresentative of the community, not bound by social obligations to it, or have no long-term interest in its welfare. That shift in power inevitably gives rise to a new set of political, economic and social structures. The society is reorganised as old allegiances and relationships are transformed. And it is this reorganisation of society, rather than the polling booth, which determines the extent to which people are empowered or disempowered.

In historical terms, the single most powerful force behind disempowerment has been the emergence of the market economy. As the vernacular economy has given way to the formal economy, economic activities which were previously embedded in social relations have become monetised: wages and cash, rather than cooperation and mutual dependence, become the basis for ensuring livelihoods. Subtle social bonds that in many instances enabled the community to act in the collective interest are undermined by new relationships that are more antagonistic and divisive. Similarly, common resources such as land and water have become commoditised and access to them is determined not by social relationships but by the market. Increasingly, communities lose control over their economic activities—and with that loss of control, their power diminishes. Their lives become dependent on markets over which they have no influence, determined by decisions to which they are not party (and can never be), and by forces which neither they, nor indeed economists, fully understand.

Disempowerment is a process that feeds back on itself. As the functions previously fulfilled by the community are stripped away, so the community atrophies as a political and economic force; it no longer functions as a community precisely because it has fewer and fewer functions to perform. The community loses its cohesion and becomes unable to resist the further intrusion not only of the market but also of the bureaucracies that now move in to take over the functions it once performed—the education of its children, for example, or the care of its old and sick. Indeed, it is one of the ironies of the market that, despite the protestations of freemarketeers, it cannot itself function without the state, for it is the state that picks up the human flotsam and jetsam—the 'rubbish people'—that the market generates. It is the market that creates much of the appalling poverty that is now a feature of so many cities in the industrialised world but it is the state that must cope with the problem. And, in doing so, it causes further disempowerment, creating dependency on a bureaucracy that has a thousand different demands on its limited resources of time and money and which, by its nature, is too distant and impersonal to treat people as anything more than 'a problem'.

Corporate bureaucracies

The process of disempowerment has been accelerated through the imposition of modern development policies on the South. These policies are specifically (and openly) intended to restructure Third World societies so that they can be incorporated into the economic, political and social structures that have emerged from the industrialisation of the North. Not surprisingly, the chief agents and beneficiaries of the 'development' process have been those corporate bureaucracies (including the state) to whom the power once enjoyed by communities in the North has now passed.

This is as true for the market economies of the West as for the command economies of the Eastern bloc. As J.K. Galbraith has remarked: 'The decisive power in modern industrial society is exercised not by capital but by organisation, not by the capitalist but by the industrial bureaucrat. This is true in the Western industrial systems. It is also true in the socialist societies... For organisation—bureaucracy—is inescapable in advanced industrial technology.'

Understanding the interests—and the behaviour—of such corporate bureaucracies is vital if we are to challenge the forces that are destroying the environment and marginalising people. Set up to manufacture or sell specific goods, to provide specific expertise, to perform specific services, corporate bureaucracies are primarily concerned with promoting their own interests, perpetuating themselves and increasing their power and influence. That desire to 'keep going' generates a tunnel vision that has a dynamic of its own. Decisions are not made because they are desirable on social or ecological grounds but because they serve particular vested interests. Indeed, time and again, we find that special-purpose organisations have manipulated research, distorted cost-benefit analyses and suppressed information in order to sell products known to be harmful or to continue activities which are detrimental to the environment.

Re-empowering communities

The power enjoyed by corporate bureaucracies—and the concomitant disempowerment of local communities—is a central feature of the modern industrial state, indeed of the whole process that we call 'development'. It is the key to understanding the systematic creation of 'rubbish people' in both the North and the South, the key to understanding the destruction of our environment, and it is the key to understanding the forces that now block change. For such bureaucracies now dominate our lives—indeed, to a greater or lesser degree, we are inextricably bound up in them—and shape our future. They are also the chief obstacles to change. If nuclear power is now being favoured by governments as the 'solution' to the greenhouse effect, it is not because there are no other options open to us, but rather that those options do not fit the agendas of those corporations that dominate the energy sector. If organic agriculture has not been widely adopted, it is not because it is impracticable as an alternative but because it has been systematically 'trashed' as such by the agrochemical industry.

Seeking to reform those bureaucracies may postpone the environmental crisis, but it will do nothing to solve it. The solution will only come if we are prepared, as a movement, to confront the more fundamental issue of disempowerment—to shift power away from the bureaucracies back to the community; to take seriously the demand from those communities in the Third World and elsewhere for local people to have 'first call on their own resources'.

To do that, the Green movement cannot operate in isolation. Nor can it succeed if its campaigns do not address broader social issues. Focusing our campaigns on the issue of disempowerment may be one way forward—not only enabling the movement to build wider alliances but also helping it to counter the view that the environmental crisis is a purely technical one. In putting forward solutions, we should not merely be asking 'Will this remove a threat to the environment or promote environmental restoration?' but 'Will this solution further the interests of corporate bureaucracies or local people?' In short, 'Will it impede or encourage communities in empowering themselves or will it further marginalise them?'

For, ultimately, it is only through the direct and decisive involvement of local peoples and communities in seeking solutions to the environmental crisis that the crisis will be resolved. As Lois Gibbs, coordinator of the Citizens' Clearinghouse on Hazardous Wastes, a grassroots environmental group in the US, puts it: 'Change does not come about through slick lobbying techniques, clever research or "magic facts" but through trusting in people's common sense and willingness to act once they are aware of the issues.' Environmental groups can never match the financial power of those vested interests against which they are invariably pitted. But they have one resource whose strength, once tapped, should not be underestimated. People.

Liberation theologians have long argued that the church has little relevance unless it is prepared to address the issues of power and oppression. Environmentalists need to address the same agenda—that of 'liberation ecology'.

This piece first appeared in The Ecologist, *Vol.21, No.1, January/February 1991.*

A VILLAGE COUNCIL
OF ALL BEINGS:
Ecology, Place, and the
Awakening of Compassion

Gary Snyder

✧

O ecology, as it used to be spelled, is a scientific study of relationships, energy-transfers, mutualities, connections, and
cause and effect networks within natural systems. By virtue of
its findings, it has become a discipline that informs the world
about the danger of the breakdown of the biological world. In
a way it is to Euro-American global economic development as
anthropology used to be to colonialism. That is to say, a kind of
counter-science generated by the abuses of the development
culture (and capable of being mis-used by unscrupulous science
mercenaries in the service of the development culture). The
word 'ecological' has also come to be used to mean something
like 'environmentally conscious'.

The scientist, we are told, seeks to be objective. Objectivity is
a semi-subjective affair, and although one would aspire to see
with the distant and detached eye of a pure observer, when
looking at natural systems the observer is not only affecting the
system, he or she is inevitably *part* of it. The biological world and
its ecological interactions, is *this* world, our very own world.
Thus, ecology (with its root meaning of 'household science') is
very close to economics, with its root meaning of 'household
management'. Human beings, biology and ecology tell us, are
located completely within the sphere of nature. Social organisation, language, cultural practices, and other features that we
take to be distinguishing characteristics of the human species are
also within the larger sphere of nature.

To thus locate the human species as being so completely within 'nature' is an unsettling step in terms of the long traditions of Euro-American thought. Darwin proposed evolutionary and genetic kinship with other species. This is an idea that has been accepted intellectually, but not personally and emotionally by most people. Social Darwinism flourished for a while as a popular ideology justifying Nineteenth Century imperialism and capitalism, with an admiring emphasis on competition. The science of ecology corrects the emphasis and goes a step further. It acknowledges the competitive side of nature but also brings forward the co-evolutionary, cooperative side of interactions in living systems. Ecological science shows us that nature is not just an assembly of separate species all competing with each other for survival (an urban interpretation of the world?)—but that the organic world is made up of many communities of diverse beings in which the species all play different but essential roles. It could be seen as a village model of the world.

An ecosystem is a kind of mandala in which there are multiple relations that are all powerful and instructive. Each figure in the mandala—a little mouse or bird (or little god or demon figure) has an important position and a role to play. Although ecosystems can be described as hierarchical in terms of energy-flow, from the standpoint of the whole all of its members are equal.

But we must not sentimentalise this. A key transaction in natural systems is energy-exchange, which means the food-chains and the food-webs, which means that many living beings live by eating other beings. Our bodies—or the energy they represent—are thus continually being passed around. We are all guests at the feast, and we are also the meal! All of biological nature can be seen as an enormous *puja*, a ceremony of offering and sharing.

This intimate perception of interconnection, frailty, inevitable impermanence and pain (and the continuity of grand process, and its ultimate emptiness) is an experience that awakens the heart of compassion. It is the insight of *bodhicitta* that Shantideva wrote so eloquently of. It is the simultaneous awakening of a personal aspiration for enlightenment and a profound concern for others. Ecological science clearly throws considerable light on the fundamental questions of who we are, how we exist, and where we belong. It suggests a leap into a larger sense of self and family. It seems clear enough that a consequence of our human inter-dependence should be a social ethic of mutual respect, and commitment to solving conflict as peacefully as possible. As we know, history tells a different story. Nonetheless, we must forge on to ask the next question, how do we encourage and develop an ethic that goes beyond inter-human obligations and includes non-human nature? The last 200 years of scientific and social materialism, with some exceptions, has declared our universe to be without soul and without value except as given value by human activities. The ideology of development is solidly founded on this assumption. Although there is a tentative effort among Christians and Jews of good will to enlarge their sense of ethics to include nature (and there have been a few conferences on 'eco-Christianity') the mainstream of Euro-American spiritu-ality is decidedly human-centered.

Asian thought-systems (although not ideal) serve the natural world a little better. Chinese Taoism, the Sanâtana ('eternal') Dharma of India, and the Buddhadharma of much of the rest of Asia, all see humanity as part of nature. All living creatures are equal actors in the divine drama of Awakening. As Ladakhi scholar Tashi Rabgyas has said, the spontaneous awakening of compassion for others instantly starts one on the path of ecolog-ical ethics, as well as the path toward enlightenment. They are not two. In our contemporary world, an ethic of concern for the non-human arrives not a moment too soon. The biological health of the planet is in trouble. Many larger animals are in danger of becoming extinct, and whole ecosystems with their hundreds of thousands of little living creatures are being elimi-nated. Scientific ecology, in witness to this, has brought forth the crisis-discipline of Conservation Biology, with its focus on preserving biodiversity. Biodiversity issues now bring local people, industries, and governments into direct and passionate dialogue over issues involving fisheries, marine mammals, large rare vertebrates, obscure species of owls, the building of huge dams or road systems—as never before.

The awakening of the Mind of Compassion is a universally-known human experience, and is not created by 'Buddhism' or any other particular tradition. It is an immediate experience of great impact, and Christians, Jews, Muslims, Communists, and Capitalists will often arrive at it directly—in spite of the silence of their own religions or teachings on such matters. The experience may often be completely without obvious ethical content, a moment of leaving hard ego self behind while *just seeing*, just being, at one with some other.

Much of India and the Far East subscribes in theory at least to the basic precept of non-harming. *Ahimsa*, non-violence, harmlessness, is described as meaning 'Cause the least possible harm in every situation'. Even as we acknowledge the basic truth that every one of us lives by causing some harm, we can consciously amend our behaviour to practically reduce the amount of damage we might do, without being drawn into needless feelings of guilt.

Keeping nature and culture healthy in this complicated world calls us to a kind of political and social activism. We must study the ways to influence public policy. In the Western Hemisphere we have some large and well-organised national and international environmental organisations. They do needed work, but are inevitably living close to the centres of power, where they lobby politicians and negotiate with corporations. In consequence they do not always understand and sympathise with the situations of local people, village economies, tribal territories, or impoverished wage-workers. Many scientists and environmental workers lose track of that heart of compassion, and their memory of wild nature.

The actualisation of the spiritual and political implications of ecology—that it be more than rhetoric or ideas—must occur place by place. Nature happens, culture happens, *somewhere*. This grounding is the source of bioregional community politics. Joanna Macy and John Seed have worked with the image of a 'Council of All Beings'. The idea of a '*Village Council* of All Beings' suggests that we can get specific. Think of a village that includes the trees and birds, the sheep, goats, cows, and yaks and also the wild animals of the high pastures (ibex, argali, antelope, wild yak) as members of the community. And whose councils, in some sense, give them voice. Then to provide space. Village territories should include the distant communal pastures and the subwatersheds as well as the cultivated fields and households. When a village is dealing with government or corporation representatives it should insist that the 'locally used territory' embraces the whole local watershed. Otherwise, as we have too often seen, the government agencies or business forces manage to coopt the local hinterland as private or 'national' property, and relentlessly develop it according to an industrial model.

We need an education for the young people that gives them pride in their culture and their place, even while showing them the way into modern information pathways and introducing them to the complicated dynamics of world markets. They must become well-informed about the workings of governments, banking, and economics, those despised but essential mysteries. We need an education that places them firmly within biology but also gives a picture of human cultural affairs and accomplishments over the millennia. (There is scarcely a tribal or village culture that doesn't have some sort of music, drama, craft and story that it can be proud of when measured against the rest of the world.) We must further a spiritual education that helps children appreciate the full inter-connectedness of life and encourages a biologically informed ethic of non-harming.

All of us can be as placed and grounded as a willow tree along the streams—and also as free and fluid in the life of the whole planet as the water in the water cycle that passes through all forms and positions roughly every two million years. Our finite bodies and inevitable membership in cultures and regions must be taken as a valuable and positive condition of existence. Mind is fluid, nature is porous, and both biologically and culturally we are always fully part of the whole.

A SOCIO-ECOLOGICAL PROPOSAL TO LINK NORTH AND SOUTH

Silvio Ribeiro and Birgitta Wrenfelt

❖

When looking at North-South relations, we are immediately confronted with a basic asymmetry. The roots of this imbalance are to be found in the expansion and worldwide spread of Europe's dominion-based values through colonialism, which legalised the decimation of millions of human beings and usurped the natural resources of the South for the benefit of the North. In some cases, including that of the indigenous peoples of the American continent, the process can be accurately called genocide.

Domination and dominion

The countries of the South will never be able to duplicate the development pattern established by the North because the world system, based on domination, continuously acts to perpetuate the inferior position of the 'developing' world. However, the degree to which individuals or social groups within the South relate to this system of domination varies. Élites, the military, and large landowners, for example, relate to it very differently than their less fortunate compatriots.

For this reason, looking at the relationships between social groups and the environment requires a knowledge of which sector, class or social organisation we are referring to. Differing degrees of power within society are important: those at the top of the pyramid of power—military officers, capitalists, politicians—will have very different perspectives on the environment than those who are in the base—small farmers, wage workers, housewives. Similarly, urban points of view will be very different

from those found in rural or indigenous communities. The
inhabitants of Buenos Aires almost certainly have problems more
similar to those of Stockholm residents than to those of their
own rural countrymen.

The perspective of cultural origin is also important. Indigenous peoples, mixed descendants of slaves, colonisers or Europeans will all have different, sometimes opposing, views about
social and environmental relationships.

An illustrative example is that of Brazil and the situation in
the Amazon Basin. For the government the Amazon might be
seen primarily as a vast supply of exploitable resources. The military might focus on its strategic value, and therefore prefer to
see it populated by modernised Brazilians rather than by indigenous peoples with a doubtful commitment to the concept of
'Nation'. For many poor peasants, the Amazon may appear to
be an empty region holding the promise for a better life, or even
survival. And for indigenous people, it would be seen as home,
livelihood, the place that sustains them physically and culturally.

In sum, the man-nature relationship is closely linked to the
cultural and socioeconomic filters through which it is viewed,
as well as to the position each person holds in the social fabric.
When this fact is ignored, oversimplified positions can be reached
which ultimately do little to tackle serious global environmental
problems. Official and international agencies generally try to
reaccommodate and adapt the use of existing resources without
questioning the model of society that engendered the crisis.

Although it is absolutely urgent to solve current environmental problems, this does not mean simply treating symptoms.
What is required is both a new concept of reality (based on the
principle that human society and nature are parts of a complex
interdependent continuum) and a new social fabric (one not
based on dominion- and domination-based values.)

Environmentalism or social ecology

A perspective grounded in environmentalism usually considers
the expressions of the ecological crisis as problems of the natural
environment. Generally, it applies a somewhat stereotyped rule,

which might be defined as 'one symptom = one step to take', and which omits an analysis of the origins of the problem. From this perspective, the energy resources crisis becomes a technological problem to be solved by the use of other resources or techniques, without questioning what the energy is used for, what the real needs are, or what the production and consumption patterns are that are latent in the crisis.

Our society—and our view of reality—is atomised; we often cannot perceive the social, political and economic structures to which we belong and which we sustain. Similarly, an environmentalist view atomises the totality and reduces it to its parts or to the simple addition of them. In this manner it contributes to hiding what are the deepest roots of the problems: dominion and exploitation, both of nature and of other human beings.

In many cases, the use of isolated measures decided upon with an oversimplified perspective may even lead to the worsening of the situation in the long term. We can see an example of this in the well-intentioned European groups that promote the consumption of Third World products as a means of favouring those countries' exports. But who are the beneficiaries of those exports?

Let's take the example of a small Latin American producer who—by means of subsidised loans and other inducements—is persuaded to leave the multiple crop agriculture through which he had achieved a level of self-sufficiency. He invests in single export-oriented crops, with a better yield per hectare and the new possibility of a substantial cash income.

He has now entered the international system, and begins to depend on forces over which he has no influence. He steadily loses control over his own production. The seeds he uses are not self-produced, and to reach a maximum yield they need both artificial fertilisers and pesticides. Thus, he becomes subject to a market in which the price of seeds, agro-chemicals and final products are fixed somewhere else in the world. His competitors—large companies and moneyed landowners—have the capital that allows them to renew tools and machinery at the pace of market innovation, and to weather the ups and downs of price fluctuations. Eventually he is led to poverty.

The massive migrations of impoverished Brazilian peasants
from fertile regions of Southern Brazil to the Amazon clearly
show that producing export crops has not been a solution to
poverty in Third World countries in the past, nor is it an alternative for the future.

Towards social and ecological development

A socio-ecological perspective must view the ecological crisis
as an inherent feature of a system of domination that places
humans above nature and some humans above others. This
perspective recognises the presence of systems—natural ecosystems as well as systems created by interrelated human beings. We
can then learn to understand the totality, and to attack the
problem not in its symptoms but in its underlying causes. At the
same time we also have to understand that both the problem
and its possible solutions are processes—actions and interactions
which as a whole are transforming. To obtain constructive results,
such processes must embrace all factors and must aim at the
fulfillment of multiple needs simultaneously and synergistically.

It is necessary to seek and disseminate alternatives focused both
on the social and economic situation and the problems caused
by environmental depletion and destruction. Only the actual
participation of the people concerned can assure another kind
of development, which must be supported by the creation of solidarity and cooperative structures. This process cannot be tackled
through measures imposed from the top or from the outside.

Though we have referred only to a small part of the ongoing
crisis, this proposal is applicable to most situations. For instance,
the problem of urban concentration and its implications—local
migrations, cultural impoverishment, deteriorating slums,
psychological and social breakdown—will only be resolved if
we can envisage viable and desirable alternatives leading to self-
reliance and self-sustaining development.

A specific expression of the move towards social and ecological development are the actions undertaken by the many grassroots groups which emerged during the last decade, and which

aim at addressing various aspects of alternative development. We are referring to groups that work for the spread of appropriate technologies, about organisations that encourage organic agriculture, or communities and cooperatives that try to reorient modes of consumption and exchange towards self-reliance. Many of the structures created by these groups within the dominant system belong to the so-called 'invisible sector' and are social 'sprouts' that attempt to create radical alternatives to the current system.

Perhaps the most interesting thing about these diverse groups is that somehow they are becoming a whole. This can only be observed from the outside, since the kind of participatory organisation and the pattern of change they promote are not necessarily explicit. An analysis of their organisation and the social and individual behaviours they promote reveal that they are a positive manifestation of a new solidarity, a social and cultural fabric consistent with a radical ecological conception.

Thus, we consider it essential to form a network of experiences represented within this trend and merge them, both in the South and the North. If we can also nurture organic links between experiences and movements, we think it may be possible to plant the seed of a new international order.

Ecology, self-reliance and cooperation

Our own experience is based on the creation of the international network Framtidsjorden (Future Earth), founded in Sweden in 1987 as a result of the work of two long-standing and well-known Scandinavian movements, Friends of the Earth and The Future in Our Hands.

This network was born after much thinking on the issues put forward here, and as a response to the need to link with and support the ongoing experiences of movements for socio-ecological development in both North and South.

Future Earth has several goals:

- *To support practical projects that disseminate a pattern of development based on social ecology.*

- *To work for an alternative development model and against the existing pattern of development.* International cooperation is essential in this effort since horizontal communication and exchange are the factors that can help overcome the faulty or incomplete information which often allows the industrial way of life to make inroads. As a specific example of this, there is the case of the attempt to establish a nuclear dumpsite in Gastre, Argentina. To demonstrate its safety and security, the Argentine government emphasised that the dumpsite would be using Swedish technology proven to be fail-safe. But in Sweden, several reports had reached opposite conclusions. So, through the promotion of international communication we can learn from other Northern and Southern campaigns and counteract the distortion of official and media information.
- *To exchange information and experiences at the international level for a deeper understanding of the problems, their causes, consequences and possible solutions.* There is a need for flows of knowledge— on appropriate technologies, organisational structures, strategies—that could be useful in other parts of the world but which currently find no room in institutional channels.
- *To provide education and training in issues related to alternative and ecological development.*
- *To foster and further not only North-South but also South-South exchange of experiences.*

We know this is a limited attempt. However, we can do no less than try. There are no longer purely individual problems since we are all touched by the present development model, the world ecological crisis and its impact. And we must strive to remember that the Earth is not ours; it is just on loan from our children.

PATHS TO SUSTAINABILITY

Wiert Wiertsema

✧

As new visions, principles and strategies for development are discussed and written about around the world, the question might be asked whether we are truly addressing the proper issues. In my opinion we are not, and mainly because of the word 'development'.

Development presupposes agents that engineer change: changes in the natural environment, changes in social structures, changes in production patterns, etc. These changes are supposed to generate improvements in the quality of life of the people involved. In this sense, development is highly human-centered, often at the expense of the natural environment. At the same time, however, development creates substantial inequalities among human populations, since different groups of actors have differing degrees of access to the power required to bring about change.

The idea of development is very much a concept from the North. Its roots go back at least to the period of the Enlightenment, the intellectual foundation for the Industrial Revolution in Europe. It was then that the early colonial powers started to establish administrative systems to rule and control huge regions in the South.

The negative aspects of these endeavours have always been apparent to some. Romanticism in the last century can be seen as an early reaction to this negative side. Development demanded heavy sacrifices from the labour class in the North, as well as from the peoples of the South. Several criticisms of the capitalist development model appeared, of which the Marxist-socialist movement eventually became the most powerful. Basically, however, the idea of development as such was not widely criticised: the powerful obviously benefited from development endeavours, and it was generally believed that if certain precondiitions for development were met, development would benefit the less powerful as well. It is now clear that this belief has proved to be false.

Certainly the standard of living of large sections of humanity has risen substantially through development, but for the vast majority, living conditions are growing worse. The development of some is only possible at the expense of many others. Marxist attempts at more socially just ways of development have also proved to be a failure. With the growing environmental awareness of the last two decades, it has become increasingly clear that all models of development practised thus far result in an over-exploitation of natural resources. Many indigenous peoples and many non-human species of life have been exterminated in this process. More and more people face the immediate danger of not being able to survive the daily struggle for life. And now even the North is speaking of an ecological crisis, because of the growing awareness that here, too, survival is threatened.

The myth of 'sustainable development'
In absolute terms humanity is facing the end of opportunities to exploit the Earth, the end of opportunities to 'develop'. Of course there are still major forces which deny this fact. Given the diminishing resilience of the global ecology, these forces have become increasingly dangerous to the health of the planet. The dominant ideology of 'economic growth' remains, but at the same time the consciousness of limits to growth is rising. Out of such consciousness, new conditions have been formulated which promoters of development have to acknowledge. Development schemes which meet the new conditions are called 'sustainable development'. Basically, this new formulation merely represents minor adjustments to the old model of development. Even these adjustments will be subverted or coopted by those agencies that have the power to do so, which are agencies dominated by the North.

It is this struggle that was dominant on the agenda of UNCED in 1992. In my view, it is clear that the role of the South in all this was very limited. The experience of the recent Gulf War underscores this view.

In this sense, I feel that the theme of ecological and social development will put us too much in the camp of sustainable development. I propose that we should talk about sustainability, or more specifically about social justice and ecology.

The need for diverse visions

Turning now to the question of a new vision and principles, I must say that I do not have a vision of what the world should look like. In my assessment of history, blueprints and powerful visions have always contributed to social injustice and ecological damage. Essentially, human and natural diversity is always more complex than can be expressed in models.

It is my conviction that, particularly in the North, a lot of restraint is required. Organisations and individuals in the North have much better access to information and to opportunities for exchange of information. This results in a position of power that is easily misused despite the best intentions. In addition, it is questionable whether the information that is gathered in the North is really relevant to the agendas of local people in the South.

Yet, with the above cautions kept well in mind, a proper sense of responsibility requires that those in the North must do what they can—not to impose their ecological vision—but to allow diverse visions to flourish in both North and South.

For this to happen, we need to start seeking equilibrium, rather than growth, as the prerequisite to ecological sustainability on a global scale. In the North this means a substantial shrinking of the economy in order to create space for selective economic growth in the South. To get to a global economy accommodating equilibrium instead of growth, an intensive dialogue between different paradigms is essential. This dialogue should involve more than just economists, since, for instance, the cultural dimension of the problem could easily be underestimated. I stress dialogue, since no one model can be expected to result in the solutions we are looking for.

Domination is another fundamental aspect of current models of development. In order to attain sustainability, the NGO community stresses the importance of participatory approaches. Empowerment of people is a prerequisite for the creation of opportunities for participation. At the same time this results in tension in the NGO community. The problems facing the world are so critical and urgent that many NGOs feel they cannot afford the time needed to promote participation from common people.

How can we maintain the principle of solidarity within a varied global community of NGOs and people's organisations? Again, more energy has to be invested in the establishment and fostering of dialogue between different schools of thought and representatives of different strategies. In addition to this, much more work has to be done on the establishment and improvement of lines of communication between local levels—from individual households all the way up to the global.

Alternatives to the global economy

Another characteristic of current development models is the integration of the whole world into one single system, dominated by the market economy. This process leads to the pauperisation of huge numbers of people as well as the elimination of natural variety. Real cultural and biological diversity does not lend itself to integration into one system. To counter that process of integration, de-linking has to be actively promoted. The global market economy has to be countered by promoting autonomous regional and local economic units. Linear and exploitative processes have to be replaced by cyclic processes. Indigenous forces in local economies should dominate over forces at the national or international level. And the local level must be the starting point for any development.

Obviously the issues discussed above sometimes overlap, and are certainly interlinked. There are no clear solutions yet. Solutions can only be created from the bottom up, by people who never will have the opportunity to meet in conferences or write in learned journals. Our task is to help them to speak out and to bring about their own diverse visions of the future.

V

Fostering Alternatives

BIOSPHERIC ETHICS

Edward Goldsmith

❖

How can we reverse the devastating effects of development on the Third World, and indeed on the industrialised countries themselves?

The answer is that we need to return to low energy, low resource, low pollution societies—and very quickly. Such societies must necessarily conduct their economic and indeed their political affairs on a very much smaller scale than is today the trend, which means catering to very much smaller markets. The correct unit for economic activity is clearly the family and to a lesser extent the community. It is only in this way that economic activities can satisfy social, religious and ecological needs—not merely narrow economic ones as is necessarily the case when they are fulfilled by corporations. Since humans, during 95% of their tenancy on this planet, have lived in tribal societies that conducted their economic activities in precisely this way, it seems clear that we must derive our inspiration from that experience.

My colleague Nicholas Hildyard and I studied traditional irrigation systems and wrote about them in *The Social and Environmental Effect of Large Dams*. Our research into the ways traditional peoples conducted their affairs showed us that in general they are difficult to improve on. This is clearly the case with their agricultural and horticultural and indeed pastoral practices in general. The literature on this subject is enormous and all of it tends to confirm this thesis.

The basis of the sustainable lifestyles of tribal or 'vernacular' societies was undoubtedly the observance of traditional laws which were seen to maintain the order of the Cosmos. So long as that order was maintained, then people prospered: if it was perturbed, if 'the balance of nature' was upset, then disaster inevitably followed.

The vernacular person's fundamental role in life was thus to maintain the order of the Cosmos by performing prescribed rituals, taking part in the prescribed ceremonies and in general by observing traditional laws of society. People understood this law to be a moral one, and one which applied not only to human beings but also to nature and, indeed, to the Cosmos itself.

Moral Behaviour

Father Placide Tempels in his celebrated book *Bantu Philosophy* notes: 'Moral behaviour for the Bantu is behaviour that serves to maintain the order of the Cosmos and hence that maximises human welfare. Immoral behaviour is that which reduces its order, thereby threatening human welfare...' This statement could apply equally well to vernacular societies in all parts of the world. In many of these societies, the pattern of behaviour that is judged to be ethical was referred to by a word that both denotes the order of the Cosmos and, at the same time, the 'path' or 'Way' that must be followed in order to maintain it.

Among the Ancient Greeks the word used was *Dike*, which also meant 'righteousness' or 'justice'. The Chinese *Tao* is a very similar concept which refers to the daily and yearly 'revolution of the heavens'. According to de Groot, *Tao* 'represents all that is correct, normal or right in the universe; it does indeed never deviate from its course. It consequently includes all correct and righteous dealings of men and spirits, which alone promote universal happiness and life.' All other acts, as they oppose the *Tao*, are 'incorrect, abnormal, unnatural' and they must bring 'misfortune on the bad'. The Buddhist notion of *Dharma*, the Persian *Asha* and the Vedic *Rita* are very similar concepts: all refer to the Way that human beings must follow if they are to maintain the order of the Cosmos, the only Way that is truly moral since to maintain it is to assure the welfare of the world of living things, while to diverge from it can only cause disasters like floods, droughts, epidemics and wars.

Although many tribal peoples do not appear to have formulated the notion of the Way in so explicit a manner, their notion of morality remains the same. Moral behaviour is still that which conforms to the traditional law and which, at the same time, serves to maintain the order of the Cosmos. Immoral behaviour, on the other hand, is that which is taboo. 'An act is taboo,' Roger Caillois writes, 'if it disrupts the universal order which is at once that of nature and society... as a result the Earth might no longer yield a harvest, the cattle might be struck with infertility, the stars might no longer follow their appointed course, death and disease could stalk the land.'

'Progress' (or the economic development with which it is equated) is quite clearly immoral and hence opposed to the 'Way' because it involves the systematic substitution of the technosphere or manmade world for the biosphere or natural world from which it derives its resources and to which it consigns its ever more voluminous and ever more toxic waste products. As the technosphere expands, so must the biosphere disintegrate and contract. Economic growth, in fact, is a measure of biospheric disintegration and contraction. The two processes are but different sides of the same coin. This means that the ethic of progress—in effect, the ethic of perpetual technospheric expansion—is in reality no more than an ethic of biospheric destruction. It is not an 'evolutionary ethic',' as Waddington and Huxley saw it. On the contrary, it is an anti-evolutionary ethic. It serves to sanctify the reversal of the evolutionary process.

A Programme for Change

A biospheric ethic, an ethic compatible with the ecological view of the world we live in, would be very different from that proposed by the industrialised world view. It would above all be one which enables humans to assist in the achievement of Gaia's overall goal of maintaining the biosphere's stability or homeostasis in the face of change, whereas immoral behaviour would be that which reduces Gaian homeostasis and hence disrupts the basic structure of the Cosmos.

The question now is: how can the critical transition of industrialised society to one based on a biospheric ethic be achieved?

The answer is only by the adoption of a carefully integrated programme, and we must assume, however unlikely it may be, that it will be adopted by the government of a major industrial nation. The programme, as we shall see, will have to be divided into distinct parts. These will all be initiated at the same time, though they will proceed at a different pace as they encounter different degrees of inertia. By its very nature, however, the programme would have to be stretched out over a considerable period of time. One cannot transform a society overnight in an orderly way. In addition, the programme would have to be accepted as a whole. One cannot phase out non-sustainable activities without causing all sorts of problems such as inflation and unemployment, unless at the same time one phases in, to replace them, other more sustainable ones. Nor can one phase in the latter without first phasing out the former so as to free labour and resources for this purpose.

For that reason, it is naive to suppose that a government elected for a five-year period can implement anything more than a patchwork of short-term expedients. It is essential that it obtain from the electorate a mandate to implement at least the first part of the programme. To obtain such a mandate, it must first of all make the electorate clearly aware of the extreme gravity of the global situation and hence of its own national one—which, so far, governments throughout the world have systematically played down.

If the programme is to be fully integrated it must be designed to reverse all the essential trends set in motion by the industrial process. The programme can be shown to consist of six functionally distinct stages (though it is not suggested that they actually occur in that order, since positive feedback would cause them to be constantly affecting each other).

The first stage is the development of a very specific world view, whose main features we have already briefly outlined. As Weber was the first to point out, without the current technologically-based worldview, there would probably have been no industrialisation.

A new worldview must replace it. A study of the value systems of traditional stable societies reveals that, though they may vary in many details, their basic features are very similar. In fact it can be shown that, for society to remain stable, a number of basic principles must underlie the worldview upon which is based its stable relationship with its social and physical environment. Let us briefly consider the basic principles underlying the aberrant worldview of industrialism, in order to see how they may be modified to give rise to an adaptive and hence stable social behaviour pattern.

• *Humanism*

It is essential to the worldview of industrialism that humans should not be regarded as an integral part of nature but rather as above it, and thereby largely exempt from the laws governing the behaviour of other forms of life on this planet. To justify this, industrialists can postulate a number of abstract entities whose possession by humans is supposed to distinguish them from the other, less fortunate forms of life. Thus only humans have a soul, they alone display consciousness, their behaviour is supposed to be intelligent, while that of other forms of life is said to be governed by 'blind' instinct. Only human societies are supposed to be capable of cultural behaviour. Such notions are unknown among traditional societies for whom nature is holy and cannot be disrupted without incurring the wrath of the gods. It is by desanctifying nature that it has become socially feasible to destroy it, and by sanctifying human progress in its stead that the process has been able to proceed at the present disastrous pace. For 'humanism' we must substitute 'naturalism'—respect for the natural world of which we are an integral but only a modest part.

• *Individualism*

Individualism is the notion that a person's duties are primarily to him or herself. This notion is in keeping with our total ignorance of the nature of natural systems of which we are a part: the family, the community and ecosystem, and of how they are related to each other. For an individual to be a member of a community, his or her behaviour must be subjected to the appropriate set of constraints. A community is an organisation. As such it displays order, defined as the influence of the whole over the parts. This influence is achieved by subjecting the parts to constraints which will limit their range of choices by causing them to become differentiated. 'Individualism' is another word for chaos. It is unacceptable in a stable self-regulating society, as it is in any stable self-regulating natural system. For 'individualism' we must substitute 'communitarianism'—the need to subject what may appear to be our individual interests to those of the community and the ecosystem.

• *Materialism*

Materialism is closely related to individualism. In traditional societies, people's goals are largely social. The accumulation of material goods plays no part in the strategy of their lives. Material goods only become necessary when they are required for the purpose of satisfying biological and social needs. Karl Marx was wrong when he referred to religion as the opiate of the people. People have always been religious. Religion is an essential part of their sociability, which assures the stability of their social environment. It is not religion, in fact, but materialism that is the opiate of the people.

• *Scientism*

Scientism is the notion that scientific knowledge can serve as the basis for social and ecological control. Let us not forget that there is no precedent for stable societies based on objective scientific information. Until now they have invariably been based on traditional and very subjective information designed to adapt a particular society to its specific environment, rather than all societies to all environments. It can be shown that only such cultural information satisfies basic cybernetic requirements. We must develop increasing respect for the information organised into the cultural pattern of remaining traditional societies. This is essential to the task of social decentralisation. For 'scientism', in fact, we must substitute 'culturalism'.

• *Technologism*

The notion that there is a technological solution to all our problems is a myth closely associated with scientism, since the solutions which scientific information can give rise to are technological ones. These, however, can play no part in the strategy of nature. We must develop instead a quasi-religious respect for the natural systems that make up the biosphere, whose normal functioning provides the only lasting solution to such problems.

• *Institutionalism*

Institutionalism is a myth closely related to the preceding ones. If benefits are material and technological, then one must create the optimum conditions in which they can be dispensed. Such conditions do not exist in the home, or in a vernacular community. Therefore institutions are set up to provide them. Ignorance of social and ecological cybernetics leads to the essential self-regulating nature of natural systems being ignored, while it is assumed that their control can be more effectively assured by institutions—that is, external or asystemic controls. For institutionalism we must also substitute a respect for the self-regulating nature of natural systems—a key component of culturalism and ecologism.

• *Economism*

Economism is the notion that things must be done because they are economic—that is, so as to maximise the return on capital or on other factors of production. This notion is totally consistent with the others. If all benefits are material, technological or institutional, economic growth must be the means of maximising them and hence of best promoting human welfare. For 'economism' must be substituted 'ecologism', the notion that things must be done to satisfy not a single end but all the basic (often competing) requirements of the community and its natural environment.

All of the above notions are deeply ingrained in the industrial worldview. In order to assure the general adoption of a new, ecological worldview, reform of the educational system would be needed. It would have to become considerably more decentralised, and the curriculum would also be changed so that the accent might shift from the random accumulation of data to the acquisition of the cultural information favouring the appropriate socialisation process.

Further Stages

The second stage in the implementation of this new way forward is the development of the technology required for achieving its goals. What is required is a shift—from capital intensive industry to developing the 'appropriate' technology for decentralised living.

The third stage is the transformation of society so that instead of satisfying the requirements of the production-consumption process, it would once more be composed of people who are, above all, members of families, communities and ecosystems, and whose behaviour is basically that required to satisfy the requirements of these systems and hence of the larger system of which they are a part, the biosphere. The process will come about automatically as society is decentralised and conditions are created in favour of the restoration of the family, the community and the ecosystem, at which point economic activities will gradually become subordinated to social ones.

The fourth stage is to reverse the system of capital generation, by means of the production-consumption process. Some capital will undoubtedly be required to finance the early stages of the programme designed to prevent social and economic collapse, and to modify the infrastructure of society in such a way as to favour its decentralisation. Slowly the need for capital will be reduced as systemic resources replace asystemic ones.

The fifth stage is the reversal of the process which built up the industrialised world: by radically reducing the scale of the production process and producing goods that are ever less destructive to the natural environment.

The sixth stage is reducing the scale of technological activities to permit the restoration of the self-regulating social systems which make up the natural world, on the basis of whose normal functioning these problems could be solved.

GLOBALISATION VERSUS COMMUNITY

Helena Norberg-Hodge

✧

Society today is faced with a choice between two diverging paths. The path endorsed by government and industry leads towards an ever more globalised economy, one in which the distance between producers and consumers will continue to grow. The other path is being built from the grassroots, and leads towards strong local economies in which producer-consumer links are shortened. I believe that moving in the latter direction may be one of the best ways of solving a whole range of serious social and environmental problems, from rising rates of crime and violence to the greenhouse effect. This may sound absurdly simplistic, but it is a conviction based on long-term observations in societies at very different levels of dependence on the global economy—including heavily-industrialised America, socialist Sweden, rural Spain, and most importantly, Ladakh, a traditional culture on the Tibetan Plateau.

When I first came to Ladakh the Western macro-economy had not yet arrived, and the local economy was still rooted in its own soils. Producers and consumers were closely linked in a community-based economy. Two decades of development in Ladakh, however, have led to a number of fundamental changes, the most important of which is perhaps the fact that people are now dependent on food and energy from thousands of miles away. The effects of this increasing distance between producers and consumers are worth looking at as we consider our own future.

The path towards globalisation is dependent upon continuous government investments. It requires the building-up of a large-scale industrial infrastructure, including roads, mass communications facilities, energy installations, and schools for specialised education. Among other things, this heavily subsidised infrastructure allows goods produced on a large scale and transported long distances to be sold at artificially low prices—in many cases at lower prices than goods produced locally. In Ladakh, the Indian government is not only paying for roads, schools and energy installations, but is also bringing in subsidised food from India's breadbasket, the Punjab. Ladakh's local economy—which has provided enough food for its people for 2000 years—is now being invaded by produce from industrial farms located on the other side of the Himalayas. The food arriving in lorries by the tonne is cheaper in the local bazaar than food grown five minutes walk away. For many Ladakhis, it is no longer worthwhile to continue farming.

In Ladakh this same process is affecting not just food, but a whole range of goods, from clothes to household utensils to building materials. Imports from distant parts of India can often be produced and distributed at lower prices than goods produced locally—again, because of a heavily subsidised industrial infrastructure. The end result of all this long-distance transport of subsidised goods is that Ladakh's local economy is being steadily dismantled, and with it the local community that was once tied together by bonds of interdependence.

This trend is exacerbated by other changes that have accompanied economic development. Traditionally, children learned how to farm from relatives and neighbours; now they are put into Western-style schools that prepare them for specialised jobs in an industrial economy. In Ladakh, these jobs are very few and far between. As more and more people are pulled off the land, the number of unemployed Ladakhis competing with each other for these scarce jobs is growing exponentially. What's more, the course of the economy, once controlled locally, is increasingly dominated by distant market forces and anonymous bureaucracies. The result has been a growing insecurity and competitiveness—even leading to ethnic conflict—amongst a once secure and cooperative people. A range of related social problems has appeared almost overnight, including crime, family breakup and homelessness. And as the Ladakhis have become separated from the land, their awareness of the limits of local resources has dimmed. Pollution is on the increase, and the population is growing at unsustainable rates.

Conventional economists, of course, would dismiss these negative impacts, which are not so easily quantifiable as the monetary transactions that are the goal of economic development. They would also say that regions like the Punjab enjoy a 'comparative advantage' over Ladakh in food production, and that it makes economic sense for the Punjab to specialise in growing food while Ladakh specialises in some other product, and that each trade with the other. But when distantly produced goods are heavily subsidised, often in hidden ways, one cannot really talk about comparative advantage, or for that matter 'free markets', 'open competition in the setting of prices', or any of the other principles by which economists and planners rationalise the changes they advocate. In fact, we should instead be talking about the *unfair* advantage that industrial producers enjoy, thanks to a heavily subsidised infrastructure geared toward large-scale, centralised production.

The changes in this remote region in the Himalayas are part of the same process that has been affecting us here in the West as well, although it has been going on a lot longer and has proceeded a lot further. It is a trend that I have witnessed in Europe over the years with the expansion of the Common Market, and in America, where 'bigger' has long been assumed to be 'better'. Trillions of dollars have been spent all over the industrialised world creating superhighways and communications infrastructures that facilitate long-distance transport. Still more is being spent on highly specialised education that makes possible and promotes industrial technologies—from satellite communications to chemical- and energy-intensive agriculture. In the last decade, vast sums of taxpayers' money have been spent on research for biotechnology—with the aim of allowing food to be transported even greater distances, survive even greater doses of pesticides, and ultimately to be produced without the troublesome need for farmers. The 'unfair advantage' these many subsidies give to large-scale producers and marketers is making it all but impossible for family farmers to compete with industrial agribusinesses, for the small shopkeeper to compete with huge supermarkets, or for any small producer to compete with corporations that can inexpensively advertise and transport their goods around the world.

Large corporate producers are given further advantages by policies that promote 'free trade'. The premise underlying trade agreements like Maastricht, GATT, and NAFTA is that we will all be better off if we continue to increase the distance between producers and consumers. As a consequence, Spanish markets sell Danish butter, while Danish stores sell butter produced in France; England exports roughly as much wheat as it imports; the average pound of food in America travels 1,200 miles before it reaches the kitchen table, and the total transport distances of the ingredients in a pot of German yogurt totals over 6,000 miles—even though all are available within 50 miles. Governments around the world, without exception, are promoting an *acceleration* in these trends in the belief that throwing themselves open to economic globalisation will cure their ailing economies. Ironically, these policies undermine the economies not only of local and regional communities, but even of the nation-states that so zealously promote them. The mobility of capital today means that the comparative advantage once enjoyed by states or regions has been usurped by transnational corporations, which are in the best position to take unfair advantage of free trade and the many hidden subsidies implicit in a publically-financed industrial infrastructure. The result has been global joblessness, the erosion of community, and an acceleration in resource depletion and environmental breakdown. Meanwhile, political and economic power is being handed over to huge transnational corporations.

But as I indicated at the outset, there is an alternative path, a significant counter trend that, despite a lack of support from government or industry, continues to flourish. Throughout the world, particularly in the industrialised countries, increasing numbers of people are recognising the importance of supporting the local economy. And within this countercurrent, attempts to link farmers and consumers are of the greatest significance. Something called Community Supported Agriculture (CSA) is sweeping the world—from Switzerland, where it first started 25 years ago, to Japan where the movement affects many thousands of people. Without support from above, people at the grassroots are taking the CSA idea and succeeding. In America, where all but 2% of the population has already been pulled off the land, the number of CSAs climbed from only two in 1986 to 200 in 1992, and is closer to 600 today. Significantly, in a country where small farmers linked to the industrial system continue to fail every year at an alarming rate, not a single CSA in the US has failed for economic reasons.

Bringing producers and consumers closer together has an amazing number of positive implications. Perhaps the most significant of these is that we thereby rebuild real community. Community is based on close connections between people, and an understanding of their dependence on one another. As we all can see when we visit a small shop in a village, people know each other and talk to one another. Nearby farmers that sell to the shop—and know the people who will be buying their produce—are far less likely to put toxic chemicals on their crops. Conversely, people who know the person who grows their food are more apt to help him out in difficult times, as did the CSA group in Kentucky that helped their farmer get his harvest in before an early frost.

The stronger sense of community that stems from shorter producer-consumer links in turn has important psychological benefits. My own experience in Ladakh, as well as research here in the West, makes it clear that the rise in crime, violence, depression, even divorce, is to a very great extent a consequence of the breakdown of community. Conversely, children growing up with a sense of connection to their place on the earth and to others around them—in other words, children who are imbedded in a community—grow up with a stronger sense of self-esteem and healthier identities.

Environmentally, the benefits of CSAs are enormous. The forces within the industrial system that pressure farmers to practise monocropping are reversed, since consumers need a range of products, unlike the large scale food distributors that demand large supplies of one crop. Meeting the needs of their members thus leads CSA farms to grow a wide variety of produce, and results in an increase in biodiversity. And almost all consumers that have the opportunity to communicate directly with the farmers that produce their food make it clear that they prefer a reduction in the amount of chemicals in their food— again turning market pressures towards practices that benefit rather than harm the environment. Vegetable varieties can also be chosen for their suitability to local conditions and for their taste and nutritional value, rather than their ability to withstand the rigors of long distance transport or their conformity to super-market standards—the cucumbers need not be perfectly straight, nor the apples perfectly round. The absence of packaging means a significant reduction in the huge amount of non-reusable, non-biodegradable waste that is daily thrown into waste dumps all over the world. Meanwhile, the shorter transport distance means a reduction in the use of fossil fuels, less pollution, and lowered amounts of greenhouse gases released to the atmosphere.

The Community Supported Agriculture movement has provided real grassroots momentum for shorter producer-consumer links. But lasting progress will require changes at the policy level as well. The unfair advantage now given to large-scale producers and marketers continues to threaten the success of all kinds of enterprises and inititative—including CSAs. For national economies and local communities to flourish we need to rethink 'free trade' policies that favour transnational corpo-rate producers, and instead aim at a better balance between long-distance trade and local, regional and national production. Most importantly we need to lobby for energy taxation on the produc-tion and transportation of goods. We should also critically ques-tion further direct and indirect subsidies for transport infra-structures and large-scale corporate production. We need to oppose government support of biotechnology and other envi-ronmentally risky, job-destroying technologies. Finally, we have to actively promote shorter links between producers and consumers—a process we can start today by publicising the incredible social and environmental benefits of CSAs. We can honestly tell people that eating fresh, delicious food may be one of the most effective ways of saving the world!

NORWEGIANS AGAINST THE COMMON MARKET

NØFF (The Norwegian Ecopolitical Community Against the European Community)

❖

NØFF is a coalition of Norwegian citizens which opposed Norway's integration into the European Community (EC). Our argument was based on an awareness of the connection between economic development and social and environmental destruction all over the world. Thus, despite the fact that *economic* indicators continue to show positive growth, all environmental indicators point to increasing degradation. From global warming to the depletion of the ozone to the toxification of the environment, examples of the destruction of nature and human society could go on and on. It is clear that something must change if these crises are to abate.

The EC: Focus on the Economic

The EC pretends to be able to solve environmental problems, but it has been clear since the initial discussions on Norway's membership in 1972 that the EC is part of the system *causing* these problems. As Sicco Monsolt pointed out at that time, the entire impetus for the EC has come from industrial capital; it was unlikely then and is unlikely now that the course of the EC would ever deviate from that which preserves the present economic system. As we indicated in 1972, this system has four basic characteristics:

- Progress is measured by the amount of industrial production
- Competition between individuals is promoted in all areas of the economy and society

- Competition for resources and markets necessitates continual technological innovation
- The only values and interests that are considered are those that can be reduced to monetary value.

The economic nature of the union is evident from the EC's stated goals of providing steady and stable economic growth through the free flow of goods, services, labour and capital among member states. Initially there was a political/ideological component to the EC (the dream of a peaceful united Europe), but after the economic stagnation of the 1970s, economic considerations took priority. Large corporations exerted heavy pressure for a free internal market and eventually it was decided that the economic union should be completed by 1993. It is clear that economic rather than political interests were also behind the decision to change from decision-making based on consensus to decision-making by a qualified majority of members.

The EC fails to address the urgent needs of the planet in a number of key areas:

The environment
When the decision was made to move towards a free internal market, the European Commission set up a task force to explore the environmental impacts of the initiative. The task force's report concluded that the union would have many negative effects on the environment, but that the connection between economic growth and environmental destruction could be broken if certain policies were implemented. Specifically, the report recommended preventive environmental standards, decentralised political decision-making, an efficient system of laws, and a least-cost approach to meeting environmental standards.

The new political union favours environmental policies that rely to a large extent on pollution fees, consumer education, and voluntary compliance from the private sector. It seems unlikely that such policies can bring about the necessary reform of the economic system.

The task force report, and the 'Dublin Declaration' which

followed, seemed to be saying that 'economic growth will solve the problems created by economic growth'. Implicit in the analysis is the assumption that environmental problems are a phase in the economic system that will soon pass. History gives no indication that this will be so. For example, the EC has had policies aimed at controlling pollution in place for the last twenty years, yet they have failed to contain the problem despite their huge cost.

The EC free market is in fact likely to exacerbate current environmental problems. The volume of goods and people transported will increase, leading to an increase in the use of fossil fuels. This means higher carbon dioxide emissions (which contribute to global climate change) and more nitrogen oxides, ozone, and carbon monoxide (which will cause a further deterioration in Europe's air quality).

Energy

Reduction in energy consumption is not an official EC policy, and it is likely that economic pressures will work against environmentalists' calls for energy conservation and the development of alternative energy sources.

The EC has been supportive of nuclear power, and France—an influential member—actively promotes its export. The negative environmental impacts of nuclear power are already well established; yet EC funds are being spent on the breeder reactor, an extremely dangerous form of nuclear power.

Biotechnology

The imperatives of free trade and economic growth are likely to override environmentalists' concerns about the safety of genetic manipulation and the new biotechnologies. Current EC policy considers genetically engineered organisms to be ordinary goods. Once an organism has been approved for use by the EC Commission, no member country can restrict its use. The EC regulations also allow for the patenting of any life form that has not been previously described. This means that 90% of genetic resources are patentable. Any ethical or environmental objections can be overridden by a simple majority of member countries.

Local determination of standards

Norwegians have supported much stricter standards on toxic substances, carcinogens and air emissions than other countries. They have been against the use of nuclear power and are generally in favour of strict controls on biotechnology. If Norway had joined the EC, many of these regulations could have been challenged as restrictions to trade and competition.

Although there are supposedly mechanisms that allow member countries to maintain standards stricter than the Community norms, rulings from the EC judiciary have consistently favoured trade over environmental considerations. Even if Norway had continued to enforce stronger standards on production within Norway, it would have had to accept imports produced under less restrictive EC rules. In cases involving the validity of stricter regional standards, the courts have required a level of 'objective evidence' of environmental change that is very difficult to meet. When there is uncertainty—as happens in the majority of cases—the courts have ruled against stronger regulations.

Some have argued that Norway would be a force for positive change from within the EC. We felt that Norway is freer to set its own course and is more effective in influencing others by staying outside the EC. The influence of a small country like Norway over the EC is insignificant compared to the influence of the EC over its small members. Decision-making is centralised in Brussels, where special-interest lobbyists have a tremendous effect on policies. Of the 3,000 lobbyists in Brussels, only 10 are concerned with the environment.

The Need for Decentralisation

Many people think that the best hope for solving environmental problems lies with supranational bodies. Although international agreement is clearly necessary in many areas, the experience of the United Nations makes it clear that real decentralisation is what is required. The United Nations is largely dependent on the superpowers, and it is similarly the large countries within the EC who exert the strongest influence. The impetus for the most important environmental agreements both within the EC and internationally has come from outside the supranational organisations.

In our view, solving the current crises will require the engagement of people at a local level. Promotion of the idea that environmental problems are global matters that can only be solved by big centralised bodies reduces people's motivation and leads to inaction.

Truly sustainable development must promote long-term ecological stability as well as meet the needs of all people—in North and South, today and in the future. The situation of the poor people of the world must be improved, but this does not mean we need faster economic growth as defined by GNP, which we know undermines sustainability. What is needed is a fundamental restructuring of the way we look at the world and our place within it. Nature is now viewed as a source of resources—little more than a machine that humans can control by virtue of their technological prowess.

We need to adopt a much more eco-philosophical view which sees nature and human society as complex, interactive, organic structures. A sustainable future for the Earth would likely be based on diverse small-scale societies, each with their own traditions, knowledge and sense of identity linked to a particular history and place. The EC vision of promoting easy movement of goods, services, labour and capital is obviously antagonistic to this vision.

Norway, however, could easily move toward this definition of sustainability. Norway's geography has fostered a decentralised development pattern where a large part of the population has been engaged in a diversity of activities—including farming, fishing, trade and handicrafts. Modernisation has already brought many changes and, for the present, there may well be closer cooperation with the rest of Europe. Given the current environmental and social crises, however, Norway is better off pursuing the security that comes from self-sufficiency, the flexibility and adaptability that comes from a society based on small-scale economies, and the justice that results from limiting monetary flows within the community rather than promoting them. This vision is clearly best pursued outside the EC.

In light of all of the above, NØFF members issued the following statement during the debate leading to Norway's historic 'No' vote on EC membership:

Manifesto

The world is experiencing an accelerating ecological and social crisis on a global scale. This crisis manifests itself as pollution of the soil, air and water, as a reduction of species diversity, and as a dissolution of local societies and cultures all over the globe.

As these expressions of crisis have spiraled, market liberalism has steadily increased its global influence. Yet the causal relationship between the two is only rarely focused upon. The Norwegian resistance to joining the European Community (EC) pointed out this relationship in 1972. Since that time, the symptoms of crisis have grown stronger and the causal interconnections have become far more conspicuous.

The causes of crisis are, first and foremost, to be found in the Western competitive/industrial system, a system that demands an unrestrained market. It builds upon a worldview in which nature and human society are machine-like, not living organic forms. It is this system that now seeks to fortify its positions through the establishment of the EC internal market and the European Economic Space (EES).

The EC has, step by step, been built up as a response to three main challenges:

- *The danger of violent conflict between the Western European powers.* The EC aims to permanently remove this threat by using economic growth as a self-propelling, dynamic force to fuse these powers.

- *The threat that the US and Japan will outcompete Western Europe in European and world markets.* This is to be countered with a European system based on the same principles, which would prevail by virtue of its greater numbers of specialists, producers and consumers.
- *The host of increasing environmental and social problems resulting from a system based on limitless economic growth.* These problems are to be solved by repair activities coordinated by supranational bodies.

The EC's next goal is the establishment of a monetary and political union. This is the logical end-phase of the EC's attempt to cope with the above three challenges. But these activities avoid going to the roots of the environmental and social crisis, thereby continuing to allow more room for destructive forces to operate.

In this sense, environmental activities are reduced to routines for the maintenance of the growth machine so that its global competitive strength is kept up in a more streamlined fashion. The EC's kind of supra-nationality therefore postpones any real solutions and allows global devastation to reach a higher level before a breakthrough of change. This supra-nationality effectively blocks all considerations affecting society's development except those favouring free competition.

We *do* need international cooperation, but it must be cooperation in which the priority is to stop and reverse the eco-social crisis. It has been our constant experience that the kind of international cooperation needed stands in direct conflict with the gigantic push to liberalise world trade. To choose sides in this conflict is our most important political task in the nineties. The necessity of a fundamental departure from the politics of growth is now so urgent that it can no longer be postponed.

The Norwegian Ecopolitical Community Against the European Community (NØFF) opposes Norway's EC membership for the above reasons, and opposes limits on Norway's freedom to cooperate internationally for a basic change of course. Norway, together with other independent countries, has to play an active role as a counterforce to systems like the EC. We must renew our efforts to build a world order on the basis of solidarity with the Third World and with the Earth, and in direct opposition to the order of market liberalism. Norway's aim must be to attack the root causes of the eco-social crisis, at home as well as in the international arena.

Contributors to this article include the following members of NØFF: Gunnar Albun, Torstein Eckoff, Jon Grodal, Karl Georg Hoyer, Sigmund Kvaløy, Arne Naess, Heidi Sorensen, and Arne Vinje. Their arguments are elaborated further in their book Supermarked Eller Felles Framtid *(Cappelen, 1991).*

ALTERNATIVES TO MONOCULTURE:
A Bioregional Perspective

Stephanie Mills

✧

In America after the First World War there was a popular song titled *How You Gonna Keep 'Em Down on the Farm After They've Seen Paree* [Paris]? It presumably was commenting on the uprooting of young men by the military, which shipped them across the Atlantic Ocean to theatres of war. Those fortunate enough to return to the countryside alive and unmaimed were nevertheless permanently unsettled by experiences which had carried them to foreign capitals, and many of them abandoned rural life for cities.

Today in the South, impressions of the world outside of village life are having a similar effect on young people. The old song's refrain, however, might have to be updated to "...After They've Seen MTV." And although there is no single, sufficiently compelling way to 'keep 'em down on the farm', there is a diverse, interconnected array of reasons to cleave to traditional, land-based ways of life: reasons such as greater psychological health, environmental stability, moral correctness, indigenous peoples' right to self-determination and sovereignty, and the value of cultural diversity to the whole human community. Some of these reasons will be more compelling to some individuals than others, but taken altogether, they begin to be persuasive.

For people in the South, it may also help to know that there is a counterculture in the West, a significant minority that is trying to arrive at the kind of wholesome, frugal, organic and satisfying way of life that has been the birthright of traditional peoples for centuries. One lively and promising part of this counterculture, a movement which is deeply informed by ecological awareness and environmental concern, is 'bioregionalism'. Less than twenty years old, bioregionalism is increasingly influential in various localities throughout North America. This is most likely because its social visions *follow*, rather than *precede* the realities decreed by landforms, waterflows, climate, and natural communities.

While an understanding of—and respect for—the life of the locale might seem to make obvious the path to sustainable culture and livelihood, arriving at those activities and practices consciously is supremely difficult. This is not solely because of the relentless consumerist brainwashing inflicted during the late Twentieth Century. The infrastructure and economies of the United States and other parts of the industrialised world are completely antithetical to household and community self-reliance and ecological health; to live apart from these systems requires maximum creative effort, and a certain amount of political work as well. Bioregionalism is, in one sense, a politics of resistance, remarkable in face of the momentum of materialism, exploitation, and dependency.

This terrible momentum is why, when there still are living cultures that allow human communities to enjoy an equilibrium with their surroundings, cultures whose people are given to cooperation as a matter of sensible custom rather than strenuous utopian design, such innocent cultures must be understood as sources of wisdom and truth for the whole, lately-confused, human family. It would be a black day on planet Earth to be bereft of the possibilities that such cultures represent. People in the West who are endeavouring to re-inhabit their life places need to know that a simple, hearty, joyous life is not an idealistic fantasy.

Learning from the Past

Part of the task of reinhabitation in North America is in learning the land-use history of the places where we live. By taking a retrospective view, we can learn from the mistakes made by previous inhabitants, and get an idea of the kinds of livelihoods and numbers that may be sustained in particular places. And, in the United States, which has such a very brief history as a nation, understanding the velocity of the radical transformations (and simplifications) of our landscapes can teach us that all manner of 'limitless bounty' can be exhausted.

I live on a peninsula in northwest lower Michigan, in the Great Lakes Bioregion. The knowledge that the rich and graceful hardwood forest that for some 10,000 years clothed our sandy glacial landscape could be felled completely within a period of about fifty years tells me that the desperate ignorance of my forebears was impervious to both the local gods and to simple common sense. The myth of the frontier outweighed their powers of observation. Each successive cutting has exposed more of our fragile soil to erosion, and created greater biological imbalances. Some animals, like songbirds, have become quite rare, while others, like white-tailed deer, have multiplied to excess in disturbed habitat and become a threat to the health and diversity of the forest's complex of vegetation, and to crops as well. Our farmers produce cherries and other fruit, spraying their blossoming orchards with poisons, warning passers-by not to enter for fear of contamination.

A bioregional approach to development in my home place might call for a restoration of the hardwood forests, and for people to acquire the skillful means to meet some of their basic needs by harvesting the woods. Of course to do this would mean a radical curtailment of material wants and a commitment to share tools and labour and particular abilities by way of mutual aid. It is this social dimension of the necessary change that seems most difficult to achieve, for if the landscape of the northwoods has been revolutionised, so has its human society, and these upheavals have left impediments to the attainment of true community.

Throughout North America, before the era of colonisation by Europe, lived scores and scores of Indian tribes, an array of peoples as wondrous and diverse as the continent itself. Their territories often corresponded with watersheds or other natural regions like grasslands or forests. Relations among these peoples were dynamic, and not always peaceful. Their relationship to their life-places, however, was largely harmless. For that reason, and because they held the Earth and its various creatures to be sacred family worthy of reverence and respect, the first peoples of North America have been appropriated as ancestors to the ecology movement. This late-breaking admiration, of small value to contemporary Indians, is nevertheless an improvement on the genocidal practices of the invading government.

Where I live, the United States alienated the Odawa people from their habitat by treaty provisions that required them to own and farm land individually. This was a violent departure from their allocation of hunting, fishing, and maple-sugaring rights to family groups by tradition. As a result of the imposition of this alien concept of private ownership of land, and the advent of a monetary economy, the Odawa people were, within a couple of generations, deeply impoverished. Today the mainstays of their band's economy are a gambling casino, which the tourists to our area like, and a small fishing industry, which so-called sportsmen do not. Once an integral part of this life-place, they are now a relatively powerless minority.

Along with the homesteaders who usurped Indian lands came lumberjacks who clearcut the forests. In vast areas of Michigan and Wisconsin, terrible fires followed logging. Flames set to burn slash and stumps and clear land for agriculture turned into fire storms and, in certain places, actually consumed all the organic matter in the soil, sterilising it. So there are areas within the region that a hundred years later have not regained their forest cover.

Despite the occasional holocaust, subsistence farming became the rule in the county where I live, with settlers coming from Poland and Czechoslovakia by way of the great industrial cities far to the south—Detroit, Michigan; Gary, Indiana; and Milwaukee, Wisconsin. The uncertainties of the US economy in the first half of the century had people travelling from factory to farm and back again, seeking cash and subsistence by turns. Until World War II, it was possible for a good many families here to support themselves largely by farming.

After the Second World War, technological and political

changes ripped through rural communities like whirlwinds, confounding peoples' values and habits and paving the way for the mindless, greedy, and unfulfilling consumerism that afflicts Americans everywhere. Tourism is one element of this consumerism, and represents the latest attempt to extract easy money from our life-places. Because many visitors expect to live at least as comfortably on vacation as they do at home, and to engage in forms of recreation—like golf—that can be provided only by real damage to our native vegetation and groundwater, the environmental impacts of this form of industry, while not as drastic as those of, say, chemical manufacturing, are serious. The employment that tourism provides is entirely dependent on the affluence of a small sector of the population, which is just the opposite of the bioregional idea of local independence, renewability, and community self-reliance. Competition for tourist attention can lead to grotesque projects, such as elaborate amusement parks which have no meaningful relationship to the place where they are situated.

I recount this history of my home place because it is the most important story I can know. It concerns the fate of the land where I live, and is thus supremely interesting to me—and that is as it should be. The histories of our life places form the background for our visions of the good lives we might wish for our natural communities long into the future. And histories from other places can help alert us to dangers that may be avoided, mistakes not to make.

There are lots of mistakes to learn from in the history of the occupancy of North America. Maybe the biggest one is underestimating the ultimate impacts of incremental changes. Another might be how foolish it is to regard community as peripheral, rather than central. Yet another is forgetting where food comes from, and how to produce it.

So the big question in my life place is not so different from the question in much of the South: How can we maintain resilience in the face of relentless and sophisticated pressure to 'develop'? In practice, *development* seems to mean 'making every place on Earth a resource colony greatly resembling each other'.

Looking to the Future

'Don't Mourn, Organise!' said one Joe Hill, a martyr of the union movement in the US. Hill's words echo forward through all manner of principled dissent. Meeting, conferring and clarifying our knowledge of the life of our life places, and articulating a clear statement of our desires and visions for the future of our communities are an indispensable beginning. Coming together as responsible citizens of our bioregions for discussion, study, and planning, is a powerful practice. It is not just a matter of fighting unwanted changes, but of developing consensus around positive, wholesome, humble intentions for the future. Mastering current information about the state of the bioregion and its governance is critical for this practice, but at the heart of it are love of place and commitment to sentient beings. And these two essentials can only be nurtured by culture. Based on my experience of a decade in the bioregional movement, I would say that the deliberate evolution of resonant cultural forms seems possible, and the creative effort is a lot of fun. Also, in polyglot North America, we have no choice but to start from scratch. In my region and just about all others, as I have described, our human population is now a rich but seldom harmonious mix of classes and of peoples from throughout the world. In some respects what we are aiming at is the creation—out of disparate materials and fragments of knowledge—of lifeways as coherent and enduring as the Native American cultures that were so long ago displaced.

In the West those of us who are seeking a desirable, practicable, and moral alternative to the hollow destructive life foisted on people through 'development' face the same inimical forces that today confront the 'developing world'. And we are heartened by similar values: respect for the Earth's generosity and grand cycles, delight in the virtue, devotion, and creativity of all our relations, and the belief that there can be a wild variety of good and harmless ways for human communities to shape themselves to the land. In the long run, simplicity and truth have a very great power, indeed.

THE LADAKH PROJECT:
Active Steps Towards a Sustainable Future

John Page

✧

During the thousand or more years that people have lived in Ladakh, change has been inevitable. There have been foreign influences from traders passing along the Silk Route from India to Tibet and China; there has been innovation and a growth of knowledge, and a constant co-evolution with the environment. These changes, however, occurred very slowly. The accumulated wisdom of many centuries of life on the Tibetan Plateau had resulted in a culture finely tuned to the needs of people and their environment: it was unlikely that wholesale alterations would prove beneficial.

During the last twenty years, however, the pace and scale of change in Ladakh have increased tremendously. The culture is today being radically transformed by the forces of Western-style development, a process in which societies are rapidly pushed along the road already travelled by the industrialised countries. These changes sweep away centuries of local knowledge and adaptation, seeking to make all cultures, in every environment, conform to a single Western model.

Ladakh is isolated behind high mountain passes, and lacks resources valuable to an industrial economy. As a result, the region was left largely untouched by colonialism and, until recently, development. Still largely dependent on its own resources and its own economy, Ladakh is not yet home to the numbing poverty, overpopulation, environmental stress, and ethnic conflict which have become the norm in the 'developing' world. But is this Ladakh's inevitable fate? And does the Western industrial model provide the only viable blueprint for future change in Ladakh?

Since 1978, the Ladakh Project has been answering both of these questions with an emphatic 'NO'. Along with LEDeG (the Ladakh Ecological Development Group, an indigenous organisation which the Project helped to found in 1983) we have been working to help Ladakh avoid the mistakes that have been made elsewhere—in the industrialised and industrialising countries alike. The goal is not to put a fence around Ladakh or to make a museum of its culture, but rather to improve the material standard of living without undermining the culture or the environment. Cultural change is now—as in the past—inevitable; but we are convinced that change can continue to build upon Ladakh's own resources and on the strength of its own traditions.

The Ladakh Project was founded by Helena Norberg-Hodge, a Swedish linguist who first came to Ladakh in 1975, just after the region had been opened up to tourism. Although the traditional culture was then a model of ecological balance and social harmony, the threat to the culture from modernisation was already apparent. Traditional self-reliance was being undermined by a growing dependency on subsidised imported goods and non-renewable sources of energy. The first signs of waste and pollution were appearing. Young Ladakhis, previously proud of their traditions, were beginning to imitate 'modern' Western ways, and to develop a sense of inferiority about their culture. The status of women, once roughly equal to that of men, began to suffer as the centre of the economy shifted away from the household and into the rapidly urbanising capital.

The aims of the Ladakh Project over the years have been twofold: firstly, to balance the overly rosy view of life in the West, so as to enable the Ladakhi people to see their own culture in a more positive light; and secondly, to provide support for ecologically sustainable development within Ladakh itself. To make these goals a reality, we have undertaken a wide-ranging educational and technical programme.

Counter-development

Too often it is said that people in the South are *choosing* to abandon their traditions by rushing after bluejeans, television sets, McDonald's hamburgers and other trappings of a Western lifestyle. If dislocation and exploitation follow, these are justified as the consequence of a decision freely made. But informed choice requires more than a one-sided view of what is being chosen. Thanks to a constant barrage of media and advertising images, the material successes of the West are widely known. But most people in the South do not know that side-by-side with the skyscrapers, superhighways, and conspicuous consumption there is also unemployment, poverty, homelessness, drug addiction, crime, family breakdown, teenage suicide, stress-related diseases and environmental cancers. Giving the Ladakhis a fuller picture of life in the West—including its 'dark side'—has been one of the primary goals of the Ladakh Project's educational programme.

In a similar vein, we work to debunk the 'development hoax'—that modernisation will eventually transform Ladakh into a wealthy, high-tech, ultra-modern version of New York, London or Tokyo. No matter how eagerly Ladakhis embrace conventional development, they will never attain the heights of material prosperity enjoyed by the wealthiest in America, Europe or Japan. The consumption patterns of the industrialised countries are possible only because they consume far more than their fair share of the earth's resources. And even if there *were* enough resources for every developing region to become like the rich parts of the West, global environmental problems are making it obvious that such patterns of resource use are highly destructive. The best Ladakh can hope for by following the conventional development path is to become a second-class member of the 'Third World', in which a small élite is able to prosper while the vast majority becomes increasingly powerless and poor.

At the same time, the Project lets Ladakhis know that there is a growing movement in the West that seeks a life more in touch with the natural world, that values small-scale, decentralised political and economic units and a strong sense of family and community—all of which are intrinsic to Ladakh's traditional way of life. Such 'counter-development' messages encourage the Ladakhis to maintain respect for traditional ways, and to be less eager to abandon a way of life that has worked so well for so long.

Dramas

One of the principal means of communicating these ideas in the villages is by presenting satirical plays dealing with the process of modernisation. Drama has always been one of the Ladakhis' favourite pastimes, and is an excellent way of introducing ideas for discussion in the community. The humorous element—usually involving exaggeratedly 'modern' young Ladakhis infatuated with everything Western—helps to distance people from the changes that are occurring in their lives, and provides the opportunity for a more detached view of the process of development.

One such counter-development play, entitled 'Ladakh: Look Before You Leap', concerns a Ladakhi teenager, Sonam, who desperately wants to be as modern and Western as possible. He spends most of his time at discos, drinking and smoking. He wears mirrored sunglasses and polyester shirts, and splashes himself with aftershave. His language is scattered with trendy English and Hindi words. He drinks whisky and rum, instant coffee and sugary tea. Everything Ladakhi is totally rejected. He refuses to sit on the floor, insisting on using a modular plastic chair instead; he pushes away the traditional food his mother offers him, treating her and the rest of his family with disdain; he laughs at his uncle as he sits in the corner, quietly spinning his prayer wheel.

One day, his grandfather is taken ill, and at Sonam's insistence the family calls in a Western-trained Ladakhi doctor who has recently returned from a study trip to America. Sonam is thrilled to have the opportunity to talk to someone with first-hand experience of life in the most modern country in the world. But he is in for a surprise. The things he thought were 'trendy' are not so trendy after all; and it turns out that much of what he was rejecting as backward in Ladakhi culture is actually very 'modern'.

"In fact," the doctor says, "America is nothing like you

probably imagine it. People are crying out for more contact with the land. They choose clothing with labels saying '100% natural' and 'pure wool', and it's usually the poor people that wear polyester. They have something they call 'organic, stoneground, whole wheat bread'. It's just like our bread, but it costs two or three times as much as processed white bread. There are also people who have houses made of mud which look just like ours, but they are very expensive; it's usually the poor people who have to live in cement houses. They always used to tell me, 'You're so lucky to have been born a Ladakhi, and not have all the problems which we have created here. Why are you now wanting to throw all your traditions away?'"

In another play entitled 'A Journey to New York', a young man from Ladakh named Rigzin goes to America full of high expectations about life in the West. Wearing an "I Love NY" T-shirt, Rigzin arrives starry-eyed and eager in America—the fulfillment of his dreams. There are bright lights, shops full of glamorous goods, tall buildings, fast cars.

Rigzin is proud of his new office job, where it seems there are machines to do all the work. His first day on the job, however, leaves him exhausted and stressed, despite having never left his little cubicle. He is invited home for dinner by an American friend, whose mother is glued to the television and whose father is too drunk to notice her. One of Rigzin's neighbours is a lonely grandmother whose relatives have moved far away, leaving her alone in her apartment with no company except the TV. And as time goes by, Rigzin begins to learn what every New Yorker knows: that homelessness, drug addiction and crime are features of everyday life, that almost nobody knows their neighbours, and, incredibly, passers-by won't even stop to help him when he gets mugged.

When Rigzin discovers that the company he works for makes pesticides that—having been banned as too dangerous for use in America—are to be shipped to Ladakh instead, he becomes thoroughly disgusted with the harshness of modern life and decides to return home. He now realises that the close-knit families and communities of Ladakh are more valuable than the material benefits of life in New York.

Ladakhi-language translations of these plays has been published, and they have been performed in a number of villages. Several radio plays in the same style have also been broadcast on All-India Radio Leh.

Village meetings and conferences

In addition to dramas as a means of education, we have also organised numerous meetings and seminars aimed at specific issues and problems. Some of the meetings are held in Leh, the capital, while others are held in outlying villages. A wide spectrum of Ladakhis attend: men and women, young and old, farmers, monks, modern sector employees and government officials. Many of the meetings are covered by local radio, and reports are published in Ladakhi so that more people have access to the discussions and conclusions. Meetings and seminars have been held on such issues as education, air and water pollution, agriculture, tourism, and traditional medicine.

Over the years, we have organised a number of international conferences to address certain issues in a more global light. These meetings give participants from outside Ladakh a chance to see an ecological society in action and to learn from the experiences of the Ladakhi people. At the same time, the meetings give Ladakhis an opportunity to relate local issues to a global context, and to see that the search for human-scale, ecological alternatives to Western consumerism is an important theme around the world. Recent conferences have focused on 'Buddhism and Ecology', 'Rethinking Progress', and 'The Future of Agriculture'.

Programmes for Children

It is particularly important to work closely with the younger generation in Ladakh, as they are the ones who are now being most directly influenced by the process of modernisation. They are far more impressed than their parents with the glittery appearance of life in the West, and far more eager to imitate Western ways. Peer pressure serves to take them further and further away from the traditional culture, towards the distant, modern role models portrayed on the television screen.

As a step towards countering this trend, the Project has recently produced a comic book version of the drama, ' A Journey to New York', which LEDeG is distributing throughout Ladakh. We have also produced two versions of an ecology syllabus—one suitable for children around the age of 6-7, the other for 9-10 year-olds—covering such subjects as energy, water, agriculture, waste and pollution, medicine and health, and cultural issues. They contain examples from around the world, but above all from the local ecosystem. The importance of a healthy environment is emphasised, as is the ecological soundness of Ladakh's traditional way of life. These have been translated into Ladakhi by LEDeG staff, and have been disseminated widely to teachers throughout the region.

An annual essay-writing and painting event on subjects relating to the changes brought by modernisation have been extremely popular, both with the children and their schools. The event offers the opportunity for creative expression about issues which do not normally reach the classroom. The children's perspective is always interesting and provides important insights into the thinking of Ladakhi society at large. In an essay competition on the 'The State of Ladakh's Environment in the Year 2000', for example, almost all the children referred to the 'fashionable' nature of modernisation, observing that development is often a mere imitation of the West rather than an answer to people's real needs. One student wrote, "They are only copying the foreign people. It is a disgrace for all Ladakhis," while another observed, "Fashionism will give rise to proudness and less fellow feelings." Yet another pointed out that "Fashions is not a matter of progress. It influences the Ladakhi culture directly, and makes it poor."

Campaigns

Over the years, we have organised many campaigns focusing on particularly urgent issues. The effort against the use of asbestos sheets for baking bread is one such example. Traditionally, Ladakhis baked bread by heating dough on a piece of slate above an open fire. When asbestos sheets were brought to Ladakh by the Indian Army as a construction material, many people began using the stronger and lighter asbestos instead of slate for baking. Ladakhis had no way of knowing that asbestos is an extremely dangerous carcinogenic material that is now banned for most applications in the West.

In response, the Project launched an educational campaign which warned of the dangers of this material. The local authorities were urged to ensure that the material would not be freely available. LEDeG produced a Ladakhi-language radio programme on the subject, and tourists with experience with asbestos agreed to address gatherings of local leaders. The campaign reached a broad section of Ladakhis, and the practice of baking on asbestos has already fallen out of use in most households.

A similar campaign has been undertaken on the issue of trash disposal. Traditionally, waste was virtually non-existent in Ladakh, since everything not consumed was fed to the animals or composted for use in the fields. However, plastic packaging, used batteries, wastepaper and other forms of detritus have been piling up along the streets and in streams, particularly in the capital, Leh. In 1983 the Project initiated an annual clean-up campaign to encourage people not to dispose of waste indiscriminately. It has had a major effect on Leh's environment, and now the government and other agencies have joined in. A recent battery campaign educated people about the dangerous effects of used batteries—particularly in streams, where many people now leave them. The first such campaign netted 35,000 used batteries, which were disposed of safely in the desert. Still another campaign succeeded in convincing the government to ban the use of plastic bags by merchants in the Leh bazaar.

Appropriate Technology

For centuries, Ladakhi villagers had been almost completely self-sufficient, dependent on the outside only for a few 'luxuries' like salt and tea. Villagers relied on sustainable, locally-available sources of energy: dried animal dung and scrub vegetation for the kitchen stove, animal power for agriculture and hauling, and simple water-powered mills for grinding grain. Since Ladakh was first opened up to the outside world, however, there has been a growing trend towards a dependence on modern technologies of various kinds, and a consequent reliance—with all its implications—on the importation of fossil fuels. This, of course, is the standard pattern which conventional development is imposing throughout the 'developing' world.

There are better alternatives. Certain 'appropriate' technologies maximise the use of renewable sources of energy and local materials while minimising the impact on the traditional culture and the environment. The demonstration and promotion of these technologies have played a central role in the Ladakh Project's effort to steer development in a more sustainable direction.

In purely physical terms the case for encouraging the use of solar, wind and water power could hardly be better. The sun shines for over 300 days a year, strong winds blow along the valleys, and streams or rivers run through the heart of every village. Moreover, the scattered nature of settlements and the extremely rugged terrain make providing energy from a central source particularly difficult.

Small-scale, localised energy production makes very good sense for other reasons. If the production of energy can remain as far as possible on a village level, the villages themselves will be strengthened, and the people's capacity to make a living from the land given much-needed support. At the same time, a strategy of decentralisation greatly reduces the extent to which people become dependent on the vagaries of the international money economy, and thus helps to avert some of the problems which such dependence inevitably brings.

This emphasis on renewable sources of energy in no way represents a 'second-best' option, as is sometimes believed. On the contrary, it offers the surest prospect of sustainable and equitable improvements in material standards of living. Events around the world in the last few decades have shown that a dependence on imported energy and large-scale centralised systems can cause not only untold ecological damage but also economic instability, inflation and massive debt. Cities are favoured over the countryside, leading to frighteningly high rates of urbanisation; the rich are favoured over the poor, with the effect that the gap between the two, both within countries and internationally, is growing.

The appropriate technology programme, developed by the Ladakh Project and now for the most part carried out by LEDeG, includes the following:

Solar space heating

One of the major difficulties of life in Ladakh is the bitterly cold winter. For more than half the year, temperatures stay below or close to freezing, dropping at times to as low as minus forty degrees. Traditionally, the only source of heat was the kitchen stove, which was fueled by animal dung and scrub vegetation. These stoves give out relatively little heat. Moreover, the fuels which they burn are in short supply, and have to be used sparingly.

In recent years, people in and around the capital, Leh, have begun using metal braziers, or *bokharis*, to provide additional heat. These serve to increase comfort levels considerably, but they are far from efficient, and require large quantities of wood or coke, which have to be brought in over the Himalayas.

A number of years ago, the Ladakh Project introduced some simple systems for solar space heating. They are all 'passive' systems, requiring no source of energy other than the sun. They are entirely locally built and can be retrofitted onto existing buildings. One of the most popular systems is the Trombe Wall— a simple design in which a sheets of glass cover one of the house's South-facing walls, which is painted black for better heat absorption. The mud brick wall absorbs heat during the day, and radiates it into the house at night. LEDeG has now retrofitted approximately 200 homes in Ladakh for solar heating, and there is a long waiting list for new installations.

Solar water heating

As expectations of material comfort rise, the Ladakhis are increasingly using hot or warm water for washing. The trend is accelerated enormously by the thousands of foreign tourists who visit Ladakh every year, and who expect hot showers in their hotels. To meet the new demand, fossil fuels (oil, coke and coal) are being imported from outside Ladakh in ever larger quantities.

As a means of combatting this costly dependence on pollution-spewing imported fuels, we have designed a number of water heating systems using solar energy. The systems are constructed of locally available materials at affordable cost, using the skills of LEDeG's craftsmen. They range in sophistification from a 'batch' system capable of heating 20 litres of water at a time, to a thermosyphoning system of 200 litres or more. Once again, only 'passive' systems have been used, since they are not only cheaper and less complex, but require no external energy source.

Solar cooking

In the traditional setting, food is cooked on very solid and ornate mud stoves. These stoves do not have chimneys, and the air in the kitchen can as a consequence be almost intolerably smoky—especially in winter, when all windows and doors to the outside are closed. In recent years, a new kind of stove has been introduced. It is essentially the same design as the traditional stove, but it is made of metal, and incorporates a chimney. More recently still, people have started using kerosene and LPG stoves. These are particularly common in and around Leh, where one can see long lines of people waiting to exchange their empty LPG bottles for full ones.

For many years now, the Ladakh Project and LEDeG have been experimenting with various designs of solar cookers. These have the potential not so much to replace the other methods of cooking, but to reduce their use, and thereby conserve precious fuels or—in the case of kerosene and gas—limit the dependence on imported supplies. The most successful design has different orientations for winter and summer use, making it highly efficient. The cooker also doubles as a batch water heater, since its size allows a jerrycan (commonly used in Ladakh for carrying water) to fit inside. This solar cooker design is very effective and popular, and there are over 250 now in use in Ladakh. They can be used to prepare most dishes commonly eaten in Ladakh:

stews, soups, rice, vegetables and various types of breads and biscuits.

Greenhouses

Ladakh's climate limits the growing season to as little as 90 days. While some vegetables can be stored underground or dried for use in winter, they are usually in short supply by the end of the cold season. Solar greenhouses make it possible to extend the growing season, and thereby greatly increase the availability of fresh produce the year round. The design introduced by the Ladakh Project is very simple, and requires only standard building materials—mud bricks, wood, glass, and water barrels for heat storage—all of which are available locally. Greenhouses are now very popular in Ladakh, and have been used to grow a wide variety of vegetables, including turnips, spinach, radishes and Chinese cabbage. Up to three crops can be obtained in a winter season. LEDeG has installed approximately 150 to date.

Hydraulic ram pumps

Since rainfall in Ladakh is very sparse (less than 10 cms a year), cultivation can only take place in areas where rivers or streams can provide water for irrigation. Villages are therefore located either along valleys floors or in the alluvial fans of melt-water streams. Water is brought to the fields by a complex series of canals, or *yuras*, sometimes many miles long, which rely entirely on the force of gravity. Areas of land which might otherwise be cultivated remain dry if they are above the level of the available water.

As a means of providing water to such land, and also to provide water in homes for domestic use, we have introduced the hydraulic ram pump. These pumps use the force of a relatively large supply of falling water from a stream to drive a relatively small quantity to a higher level. No outside energy is needed.

Commercial ram pumps have been available for many years, and are both effective and long-lasting. However, they are also quite expensive. We have succeeded in producing a relatively low-cost pump which can be made from standard plumbing parts by local craftsmen with only the most basic workshop facilities. These pumps produced in LEDeG's workshop in Leh are today being used to pump water for domestic use in homes and guest houses, and are widely used for watering vegetable gardens. One installation raises water some 40 metres to a mountaintop monastery.

Improved water mill

The grinding of grain was always an integral part of village life. Each village had one or more simple water-powered mills, or *rantaks*, where barley and wheat were brought for milling. In recent years, however, people have begun taking their grain into Leh, where a diesel mill grinds it many times faster than was possible with the traditional mill (and in the process heats it to a temperature at which it loses much of its nutrient value). The Ladakh Project has introduced an improved mill, which can be easily retrofitted to traditional *rantak* sites. While not reaching the same speed as a diesel-powered unit, the improved mill grinds grain much faster than traditionally, in a way which is not only much healthier than the 'modern' alternative but which serves to strengthen, rather than undermine, village life.

Micro-hydro power

The increasing centralisation of energy supply systems in Ladakh is particularly evident in the provision of electricity. With a few exceptions, villagers depend on a very small number of diesel-powered generators located in the Indus valley, or, as of 1987, a 4-megawatt hydro-electric plant on the Indus River, which was commissioned to meet the rapidly increasing demand. Such systems have a number of shortcomings, when all the 'hidden' costs are taken into account: they are generally very expensive; they are liable to pollute the environment; they cause widespread disturbance during their frequent breakdowns; they increase the level of people's dependence on forces outside their control; and they favour the urban rather than the rural sector, thus further encouraging centralisation and urbanisation.

   In order to encourage a greater decentralisation of energy

supply, and to explore ways in which as many needs as possible can continue to be met from within the village itself, we have helped small villages to install micro-hydro plants to meet their basic needs for lighting. When not being used to generate electricity, the units can be employed to run mills, saws and other equipment.

While the alternator is obtained from Delhi, the foundation assembly, turbine casing, nozzle and other parts are manufactured in the LEDeG workshop. Villagers are responsible for construction of the waterways and the plant's enclosure. They are also responsible for operation and maintenance of the station once it is commissioned by LEDeG staff. A villager is trained for six months by LEDeG to enable him to operate and maintain the station.

Local Economy

For more than a thousand years, agriculture has been the foundation of the Ladakhi economy. Despite the extreme climate, Ladakhis have been self-sufficient in virtually all of their basic needs, using entirely local and natural inputs. Ladakhis grew all the grain they needed for the year, and even an excess for brewing *chang*, the local beer. Dried peas and vegetables ensured a balanced diet year-round. Animals provided meat, fresh dairy products, and butter and cheese that could be used through the winter; animals also furnished wool for clothing, power for transport, ploughing and threshing, and dung as fuel for the kitchen stove. Human 'nightsoil', collected from composting toilets within each house, was the primary fertiliser. The diversity of crops, as well as the cold and dry environment, minimised damage from pests. All irrigation water was from snowmelt, directed into carefully constructed channels and terraced fields. Local breeds and varieties of domestic animals were hardy and well-suited to Ladakh's climate; carefully controlled grazing allowed the high alpine pastures to support them without being degraded.

Traditional diversified agriculture in Ladakh was a sustainable, balanced system that provided a more than adequate food supply for all Ladakhis. The fact that every family had surplus to trade for luxuries is a testament to its productivity.

Traditional methods are still clearly productive enough to provide for all of Ladakh's needs. In recent years, however, this ecologically sound system has been under assault from many sides. Government 'development' programmes promote and subsidise chemical fertilisers and pesticides. Hybrid seeds, which cannot regenerate themselves, are being introduced. Heavily subsidised rations of imported rice and white flour discourage local production. The status of the farmer is declining, and agriculture is becoming a 'second-best' occupation, meant for those without the qualifications needed to work in the modern sector. More and more farmers are leaving the land, and those who remain are moving away from diversified, self-reliant agriculture towards single-crop cultivation for cash. In parts of Ladakh where 'modern' methods have been practised longest, farmers have already noticed a deterioration in the quality of the soil. People are also commenting on how the vegetables grown on this soil lack the taste of the vegetables grown more naturally.

In response, we have implemented a programme aimed at restoring respect for the farmer, while exploring ecologically sound ways of building on traditional techniques. This programme includes:

- Distributing vegetable seeds as a means of increasing the variety and nutritional content of the diet, and working to reestablish such valuable traditional practices as seed exchanges;
- Encouraging the cultivation of a number of indigenous crops which are in danger of disappearing;
- Conducting agricultural research to compare the yields of organically-grown crops with those grown using chemical fertilisers;
- Holding regular farmers' meetings and seminars throughout Ladakh, which underline both the strengths of traditional methods, and the environmental, economic, and social problems associated with modern, intensive methods; we point out that that many farmers in the West are now moving away from

chemical agriculture, and that consumers there are willing to pay more for naturally grown grains and vegetables than for artificially grown ones;

• Arranging study tours within India and abroad, to enable farmers to benefit from the experience of groups and individuals around the world that are seeking to return to or maintain organic, sustainable agricultural systems.

Village employment

Around the world, conventional development encourages farmers to leave the land in search of paid employment in urban centres. While the village economy is still to a great extent alive in Ladakh, more and more villagers are abandoning their self-reliance in order to make money in Leh. This urbanising trend makes people dependent on imports for their basic needs, in an economy over which they have no control.

Our handicrafts programme is aimed at helping villagers to earn a cash income while continuing to benefit from the agricultural economy. Rather than encouraging the creation of full-time crafts professionals who will sit all day over a loom or lathe, we hope to make it possible for the great majority of Ladakhis—young and old, male and female—to supplement their livelihood through part-time work at home. Ladakhis are fortunate in having six to eight months of winter with little or no agricultural work, and much time to devote to other activities. This programme encourages people to continue to provide for their basic needs from local resources, while enabling them to earn a supplementary income with which to buy goods that are only available with cash.

Training is provided in a range of skills, from painting to embroidery, tailoring to woodcarving. With some professional assistance, products have been developed which build on traditional structures and designs. Training takes place both in Leh and in numerous village centres throughout Ladakh. Once qualified, trainees return to their villages and are assisted in setting up local cooperatives.

There is an enormous market for traditional handicrafts in

Ladakh, principally from the 15,000 or so foreign tourists that visit the area every year. As of now, the vast majority of the goods for sale in Ladakh are not made locally, but are imported from Nepal, Kashmir and other parts of India. For the time being, these tourists provide the principal market for the crafts, but in the longer term some items may be exported to crafts shops abroad.

This programme promotes an alternative economic model which preserves the self-sufficient local economy and counters the current urbanising and centralising trend. While enabling villagers to earn a cash income, it allows them to remain in their own communities and to avoid becoming overly dependent on the volatile macro-economy.

An Ecology Centre

The Centre for Ecological Development in Leh is the head-quarters for LEDeG, and provides local offices for the Ladakh Project. The Centre houses technical and crafts workshops, cultural exhibits, a library, restaurant, handicrafts shop, offices and meeting rooms. The building itself is an example of how tradi-tional architecture can be 'updated' to meet changing needs and expectations. Most of the building is solar-heated using simple passive techniques, and utilises a solar water heater. A small wind-mill provides electricity for back-up lighting. In the garden there is an array of solar cookers and food dryers and a solar green-house, all of which are in active use.

The Centre's library seeks to show the extent of the world-wide trend toward more ecological, human-scale ways of living. There are books and magazines on environmental issues in general, organic agriculture, renewable energy, and natural health-care. In addition there is a growing collection of works on Buddhism and on the culture of Ladakh and other Himalayan peoples. There are also newsletters and directories of environ-mental organizations in India and abroad.

The restaurant serves a wide range of Ladakhi foods and other dishes, organically grown and solar-cooked whenever possible.

An exhibit intended primarily for Ladakhi schoolchildren high-lights the contrast between traditional Ladakh and the urban squalor that can result from conventional Western-style development.

The Ecology Centre's engineering and technical workshop is one of the best equipped in the region, allowing LEDeG to manufacture within Ladakh major parts of even the more complex appropriate technologies.

Every year the Centre is visited by as many as 5,000 foreign tourists, whose presence serves a very important function. Their interest in ecology, solar energy, small-scale hydro power, wind energy, and organic agriculture demonstrates that a development path based on these concepts is not a second-class path, but is in fact a path that many modern, 'up-to-date' Westerners would find preferable for their own countries. Tourists are also encour-aged to communicate some of the reality of life in the West beyond the glamorous media images, and can thus serve as active participants in the much needed work of counter-development.

GRASSROOTS DEVELOPMENT IN THE ARAVALI

Aman Singh

❖

The Aravali hills, one of the oldest hill systems in the world, spreads throughout Rajasthan in Northern India. This bioregion is home to a diverse range of flora and fauna, including much of India's remaining tiger population. Over the centuries, several civilisations and many tribal groups have flourished in the arm of the Aravali. In recent years, however, water scarcity and the activities of commercial industrial interests have led to a deteriorating situation among the tribal people. Economic decline, indebtedness and emigration have been the consequences.

To counteract these problems, a youth group called Tarun Bharat Sangh was formed in 1985. This non-governmental organisation emphasises strengthening the autonomy, self-sufficiency and self-respect of Aravali's tribal villages. Sangh has combined service programmes with consciousness- and awareness-building. It encourages farmers to adopt ecological, sustainable practices, while also attempting to show villagers that traditional talents and skills can be used to improve the local economy and conserve natural resources.

These notions often run counter to the development ethic that permeates the government at almost every level in India. In 1986, for example, Meena (a tribal community of Gopalpura village) used their traditional knowledge to retain rainwater on their community land by building a *johad*, or small earthen dam. The aim was to help recharge wells, stop fertile top soil from being carried away by runoff water, irrigate crops, and green the village surroundings. The positive results led the villagers to build two more water harvesting structures. These villagers did not consult geologists when they built their *johads*, but instead used traditional knowledge and skills. Despite the success of these wells, development experts had little appreciation of them, and the government declared them 'technically unfit'. In the course of time, however, the government and its big brothers had to accept the fact that the villagers had done a very effective job of soil and water conservation. About 200 *johads*, benefiting 250 villages, have now been built in the region. But the fact remains that similar productive systems once used by traditional people everywhere are being neglected or destroyed by so-called development programmes or in the name of scientific progress.

Another example of the use of local knowledge systems is that of Deori Village, which is situated in the core area of the tiger sanctuary, Sariska National Park. These villagers live harmoniously among the forest's flora and fauna, and by conserving forest resources have been able to minimise drought in summer and floods in monsoon. Conservation is an inherited characteristic of this culture. The people have evolved a system of fines, whereby if any person should harm the flora and fauna—either by cutting or poaching—he is fined at least Rs. 11. If someone is known to have witnessed such an unlawful act without reporting it to the *Gram Sabha* (village committee), he is fined at least Rs. 21. If the individual repeats the mistake, the *Gram Sabha* exerts social pressure to bring about the necessary changes. The final remedy is to expel them from society.

So far as agricultural practices are concerned, these illiterate villagers are probably far more skilled than an agriculture graduate of any modern university, because they learn everything from practical experience. Traditional land use and occupational structures have invariably been location-specific and ecologically sound. The villagers' natural approach to farming requires immense hard work and an alertness to nature. Industrial inputs—hybrid seeds, chemical fertilisers or pesticides—are not used, except in a few villages that have been influenced by development 'experts'.

Local control
The experiences of many such villages prove that ecosystems can be effectively managed on a local level. In some cases this is

made more feasible when legal authority over the ecosystem is given to smaller, decentralised units. The small village of Seed, for example, is registered under the Rajasthan Gramdan Act 1971—a unique law which gives executive and legal power to the village *Gram Sabha*. The act allows the *Gram Sabha* to manage natural resources within the village boundary, giving it also the power to judge, penalise and prosecute. Therefore, the *Gram Sabha* of this village has full control over all the land within its boundaries. It has defined rules for each and every concern, in ways that reflect the specific needs of the village and the local ecosystem. Centralised authorities cannot understand the ecosystem in as detailed a manner as the local villagers, and so rules applied from above would not be as effective.

A similar example is that of Madalvas, a village located to the south of Sariska National Park. This village is characterised by a subsistence economy: most of what is produced is consumed locally. The *Gram Sabha* manages their ecosystem by exercising their own rules, unanimously decided by villagers, even though the Madalvas *Gram Sabha* is not registered under the Gramdan Act. In fact, the *Gram Sabha* is an informal body. It is obligatory for all households to attend the *Gram Sabha* meeting, usually held twice a month except during the harvest season, when emergency meetings are called if necessary. There is no single leader or core group that oversees the management system. Instead, all households take an active part in the working of *Sabha,* and take all decisions by consensus. The villagers have been successful in protecting their forests and grazing lands, and in conserving rainwater within the village boundary.

In some villages, the decline in traditional knowledge has been associated with the degradation of common property resources. These resources are 'owned' by the community as a whole. These resources can be natural (e.g. land, water, forest, pastures, the village bull) or social (e.g. playgrounds, meeting places, roads and paths, monasteries). Since systems of traditional knowledge are often closely linked to the careful use and maintenance of these resources, Tarun Bharat Sangh is working to educate and train village youths in common property resource management. The objective of this training is to develop an understanding, particularly among youths with questioning minds, of the importance of these resources in village life. The main emphasis is given to the traditional subsistence economy of the village.

Nowadays, a lot of studies emphasise that each rural settlement—not only in India, but globally—should have its own clearly and legally defined environment to protect, improve, care for and use. Villagers themselves have also started to rethink the 'progress' that has been foisted on them by development experts, and are beginning to seek out their traditional values. Suratgarh village, situated on the outskirts of Sariska National Park, is perhaps a typical example. Suratgarh considered itself 'progressive', and in the name of development all traditional systems were being neglected. However, they recently 'woke up' and started to redevelop their natural resources—water, pasture, forest, etc.—using their traditional skills.

There are still many communities in rural India following a traditional way of life. For them, and for rural people around the world, the sustainable paths to the future may still use guideposts from the past.

THE GREENING OF SWEDEN

Märta Fritz

❖

Sweden has definitely become greener during the last decade, at least in the sense that more and more people are becoming aware of the urgent need to decrease pollution and garbage, and to change to a more sustainable kind of development.

Politicians and industrial leaders are, as never before, saying the right words, expressing the need for reusing and recycling products. But there are also some grassroots movements which have started their own activities to promote change.

Green consumers

Although not necessarily organised as such, green consumers have, for instance, accomplished an almost total abolition of the use of chlorine-bleached paper in Sweden (but of course we still export it!). The demand for organically grown vegetables is steadily increasing. The extensive use of packaging is being criticised and many people now compost their household garbage. A new law has made municipalities responsible for sorting garbage at the source, that is, in households. Consumers have—with some success—opposed such things as the use of non-renewable resources for plastic bottles and aluminum tins, the production of batteries containing mercury or cadmium, and so on. At almost any shopping centre you can find boxes or 'igloos' for used batteries and glass, and all municipalities now collect newspapers from homes.

All these are good signs and good initiatives, but it is obvious that they are not enough. Sustainable development in its complete sense is much more that a decrease in pollution and garbage. For real sustainability, the use of non-renewable resources must stop altogether. One prerequisite for an ecologically and economically sustainable society is that it abandon the linear flow idea which dominates all production in Western culture: you take something from nature, make a product from it and throw it away after use. The approach called 'circular flow thinking' is necessary for sustainability—everything should somehow go back to nature in a useful, harmless way. What is taken out must be given back.

Eco-villages

Among the new grassroots movements are the eco-village associations: groups of people coming together with the intention of building a small village as sustainably as possible, within practical, economic and lawful limits. The initiatives were originally taken by groups of people with this common interest, but in the last year municipalities too have begun to offer land for building eco-villages. So far there are about 5 villages in existence, 20 under construction, and 50 to 60 more groups who plan to build them.

An eco-village should have low water and energy consumption and should use only materials and substances which are not harmful to nature. Energy should be taken primarily from renewable sources like sun, wind and water. The village should have its own system for taking care of all waste and garbage (to recycle or use as manure). In some of the existing eco-villages, the water closet is excluded altogether; in others, there is a specially constructed toilet that separates urine and feces, thus making composting easier.

For most people living in eco-villages, the social aspect is very important. Modern society has removed human beings from nature and from their relatives. In central Stockholm, for example, 63% of households are one-person homes! Among people who care for the environment and our common future there is now a strong reaction against this unnatural way of living.

Most eco-village groups want to have a lot of activities in common and they have all built or plan to build a central house where all inhabitants of the village can meet. Many plan to have their own day-care centre and workplaces within the village (for instance, for people who do computer work). Most believe that the people in the village should not use transportation to the same extent as most people in the West. This also means that many want to have an organic farm close by, from which the villagers can buy their food. In this way they know who grows their food and how it is done, while transportation is minimised.

This last point touches another basic eco-village characteristic: residents want to know as much as possible about the functioning of the village as a whole—where the water comes from (many want to drill their own wells); where the electricity comes from and how the power plant works; where their food comes from; how the animals who feed and clothe them are treated; where the waste and garbage goes; what the local ecosystem is composed of; and so on. It is important that individuals have an overview of how their basic needs are met, in order to give them a feeling of control, to create an understanding of the possibility of making changes when something is wrong, and to avoid feelings of powerlessness and meaninglessness.

There are as yet no agreed-upon rules for the definition of an eco-village. The ones in existence in Sweden so far vary in the extent to which they are truly ecological. But the mere fact that many people want and strive to live in more ecologically sound ways is positive, as is the increasing interest from the government and their new willingness to change restrictive rules and laws.

The next step in eco-village development is to look at converting existing buildings and the big apartment houses in towns and cities. In urban areas, there has already been a slight increase in the use of solar heating and greenhouses.

The future 'perfect' eco-village—connected with one or more organic farms—would be one that is altogether ecologically sound, and in which a bartering system is built up parallel to the larger economic system, thus decreasing vulnerability to external crises of various kinds. Of course the villages should not become isolated islands (a criticism that is sometimes raised). In the computer age, this seems an impossibility; and of course people in the villages want to connect with other people, as humans have always wanted to do whenever possible.

Kernelfarms

Another grassroots movement growing in Sweden is so-called kernelfarms. Kernelfarms all have several characteristics in common:

- *openness*—the farm is open to visitors who want to learn practical and theoretical knowledge about agriculture;
- *carefulness*—kernelfarms are run without chemical fertilisers or pesticides, and every possible concern is given to life in all its forms;
- *farsightedness*—kernelfarms have a long-term perspective, which manifests itself in everything from protection for threatened species to plans for managing water and energy in times of crisis.

There now exists a network of around 80 kernelfarms in Sweden and a support organisation. An expanded network of kernelfarms will create a good civil defense for human beings, animals, nature and culture. There are also farms in Norway, Denmark and Finland, which strive for the same goals.

Eco-municipalities

The overall aim of the Nordic eco-municipalities is to achieve economically and ecologically sustainable societies. Groups in Sweden, Denmark, Norway and Finland are all working in this direction. In Sweden an eco-municipality project begun in 1990 now involves twelve municipalities, and more are joining all the time.

The aim of an eco-municipality project is to support municipalities in their planning for a sustainable society. In a pre-study, the municipalities define the concept with regard to their own local situation, design an action plan for change, spread and anchor the ecological perspective in the municipality in order to broaden the discussion in the pre-study, and develop pilot projects.

The pre-study is carried out with a 'grassroots perspective' as the guiding principle. It is for the municipalities to decide how the study will be managed.

The eco-municipality project was initiated in Sweden by the Nation Board for Sparsely Populated Areas, and was inspired by Overtornea, in the northernmost end of the country, which in 1983 declared itself an eco-municipality.

The basic view of the project grows from circular flow thinking—that is, that societies' supply systems must reflect the circular flow of nature. This thinking is based on the fact that resources in nature are never destroyed; they are used and transformed into materials usable in other natural process. These processes in turn transform the materials back into new resources, allowing the cycle to continue indefinitely. Modern industrial systems have substituted a linear scheme for nature's cyclical one, using—and essentially destroying—vast parts of the global ecosystem. In a sustainable society, the circle must again become whole.

GLOBAL PERSPECTIVES ON LOCAL ACTION:
Notes from Stocken, Sweden

Erni and Ola Friholt

❖

This paper will deal with the economic and cultural development of Stocken, a small fishing village on the west coast of Sweden. There will also be some discussion of similar developments elsewhere.

Stocken today has 150 inhabitants. Employment is mostly found in various places nearby and up to 50 kilometres away. The village is situated on an island called Orust, with a total population of 1200. Its traditional craft is ship- and boat-building (nowadays production of luxury boats for the world market). Until 1970, Orust's coastal villages were characterised by fishing and related activities. Today, most of these small villages are reduced to places where people spend their free hours and sleep. About ninety percent of the traditional houses are now owned by outsiders and used as summer houses for about seven weeks of the year. The remaining time they are mostly empty and unlit. In summer there is hectic vacation life, with fast noisy boats, hundreds of cars parked everywhere, various forms of watersports, fishing, and so on. In June, the lifestyle of the cities is stamped into this tiny fishing village, but by the middle of August it is totally switched off in a few hours. Then the tiny signs of permanent local life again emerge.

Eroding a locally-based economy
Stocken's economy used to be based on local activities such as fishing, agriculture, and handicrafts. Now the village has become totally dependent on outside employment, outside investments and outside tastes and cultural influences, which tend to

annihilate its original character. In this process government policies play a significant role—for example, in taking away regulations which supported fishing and small-scale agriculture; fiscal
policies; finance policies, especially those that reduce availability
of loans for traditional economic activities. These government
policies fit into the global economy of markets, transnational
companies, cash-book economics, profits from cheap imports
from Third World countries, and so on.

From the point of view of the city-based majority of the
Swedish population, a dying, lethargic countryside has few values
to really enjoy. And it seems a bit difficult to maintain supremacy
over a wilderness which is used only for recreation or as a mine
for resources for city people. So for the past three years there has
been a city-based campaign for revival of the countryside. The
conditions for revival were the same as those prevalent in cities,
that is, 'progress' through modernisation of economic life, making
money available for profitable investments, and so on. To many
of us living in the countryside this is of course only an example
of curing one manifest evil with much more of the same evil.

Obstacles to a sustainable future

The old cosy fishing village is a possible tourist asset which could
be exploited by the local people. But what are the actual facts
in the matter? As mentioned earlier, this old place has gradually
become a summer resort that is next to lifeless in winter. For
this kind of alternating life the whole place has been dressed
with asphalt. Parking places have been prepared all over. The
seashore line was once a slowly sloping series of polished rocks.
It is now covered under one or two metres of dumped stones,
flattened on top and dressed with asphalt painted with the city
pattern of white parking squares. From the sailor's point of view
it is not a good place to be shipwrecked. And there is absolutely
no poetry at all in the meeting between the small glittering waves
and the messy boulder shore. In the inner end of the bay there
were once meadows all along the shore, partly waterlogged,
partly flowery, with various birds. The flat land slowed the flow
of rainwater into the bay and fostered abundant fish nurseries in

the shallows. Eels could be caught there by the thousand every year.

Today the flat land has been provided with drainage, which allows for three months' camping for hundreds of vehicles yearly. The acidic rains go straight into the shallow waters, and there are no more fish to see or catch. The camping enterprise has gone through various stages of fraudulent bankruptcies, used by profit-makers from the city to wipe out their profits in other places and thus avoid taxation. Now it is at least partly in the hands of people who live on our island, if not in our village. There are also electronic gambling machines, a miniature golf-course and a hamburger stand. And on neighbouring agricultural lands there are plans to make nine-hole regular golf links.

In the opposite outskirt of Stocken there is a marina project that has been going on for many years. Owners come and go, making messes and hidden profits and always getting away with failing to protect the environment. They have dumped clay and constructed a new road through a settlement where many small children live. The shallow fish nurseries have been disturbed and the three hundred or so boats that come every summer each produce about two kilogrammes of poisonous paint. This paint peels off in small dots as soon as any seaweed tries to grow on it. Part of the paint also spontaneously falls off or dissolves in the salt water of the bay. Poison is killing the bottom fauna and the nearby waters.

The people of Stocken are also changing. Some are willing to adapt, but others feel that their identity and lifestyle are seriously threatened. Gradually, the permanent inhabitants are beginning to feel that they are merely parts of a museum fishing village owned and ruled by city people on holiday. Even the sea birds are scared away. The place has become even more silent than before.

This intermediary stage provided the impetus for the campaign for revival of the countryside. It all started with the early steps towards modernisation. For instance, water toilets were installed, together with sewage treatment. The former cyclic process of dumping the wastes on farm compost heaps has been broken. Today, fishing inside our archipelago is almost nil, and the burning of oil for heating houses and the use of tap water are all contributing to the ongoing eradication of self-sufficiency in a reasonable local economy.

Local solutions
The situation in Stocken is similar to many places all over the world. Having studied this issue for more than two decades, we are now more convinced than ever that the traditional labour movement has no solution to these developments and their consequences. Two of our basic assumptions are the following:

- We have to adjust to ecological facts, and whatever changes we make in this direction adds to the general consciousness of all.
- No authority or social group that has economic and political power will ever give anything up in order to make it possible for all to live in reasonable conditions, even if providing those conditions does not lead to resource competition with those of the élite class.

So, evidently, we have to do it ourselves. And we have to start from the beginning, thinking in terms of new education programmes that are promoted outside the officially supported channels.

In Stocken, the authors have tried to practise a sustainable life for many years. We don't own a car and instead cycle, walk or go by public transportation. We have reduced our personal consumption to a reasonable level (like that of Sweden in the 'fifties), first by not buying all the electronic gadgets available. Our personal circumstances were such that we could avoid borrowing money for our private projects, partly because there were very few. Our old house included an antique boathouse and a concrete platform on the seaside, which we decided to use as a café. We bake all the cakes ourselves from raw materials as far as possible purchased and produced locally. We make ice cream according to old natural recipes with three ingredients instead of forty-two, use eggs from 'free-range' hens, flour from farmers who use ecological, organic methods, and we serve coffee, tea and cocoa from self-managing local cooperatives in Mexico, Tanzania, Zimbabwe, Vietnam, Bolivia, and other countries. We try to propagate boycotting and avoiding the transnational corporations and local blood-suckers.

We cultivate part of our own vegetables, potatoes, berries, flowers, and so on, and we pick wild berries and mushrooms. We have tried to reduce our oil consumption by using solar energy captured by home-made solar panels. And there are waterlogged fields nearby where we hope to grow biomass for fuel.

In our boathouse we exhibit oil paintings and sell books. We invite singers and musicians and theatre groups to preach our message. Our general beliefs are based on ideas of Marx, Mao, Gandhi, Nyerere, Sankara and others. But we are convince that there must be a diversity of ideas—unique and different for each village.

We produce cheaply-printed handouts about ecologically-based alternatives for café guests. These handouts say that our society is a consumer society based on extortion of work and goods from people elsewhere, especially on other continents. And that we cannot save people elsewhere; we can only save ourselves, but in a way that agrees with our own knowledge and conscience. We are helping our local villages to survive and resist world market powers, because we favour our own local production. We can gradually reduce the power of the big companies and support the idea of a future small-scale economy. Groups working for these same goals can be be found in Tanzania, Mexico, Peru, and Nicaragua. By reducing the pressure on their raw materials and buying what we need from their cooperative local efforts we support the same processes there.

There is a risk that those who begin to analyse the problems will get stuck in the traditional power and counter-power system, thinking in terms of labour unions, and so on. Not that labour unions are not needed and useful, but they always halt at the level of power over production. To get more rights within the existing power structure doesn't solve the problem. Even land reform may fail for lack of context.

As consumers, we have a lot of power. By changing our consumption patterns, we support the building of new structures. Even if we fail in creating new structures we have at least added to the general level of awareness. We believe that there will be historical moments when certain ideas catch on and spread through action. And we hope to prove it in a small way.

Notes on the Contributors

❖

Dr Nsekuye Bizimana is a specialist in veterinary medicine. Born in Rwanda, he is the author of *White Paradise, Hell for Africa*, which draws heavily on his experiences of life in Germany, where he has been living since 1970. He is currently working on a research project on traditional veterinary medicine in central Africa. *Address:* c/o Edition Humana, Grainauer Strasse 13, 1000 Berlin 30, Germany.

Erni and Ola Friholt, former teachers turned freelance journalists and lecturers, are concerned with exploring alternatives beyond the modern industrial paradigm. They also work as editors for peace and environmental groups in Sweden. *Address:* Stocken 2127, 440 80 Ellos, Sweden.

Märta Fritz is a freelance journalist and editor. She was formerly a lecturer in genetics at the University of Stockholm and editor-in-chief of *Alternatvet*, the weekly newspaper of the Swedish Green Party. *Address:* Ringparken 5, 131 50 Saltsjö-Duvnäs, Sweden.

Peter Goering is the former Research Coordinator for the International Society for Ecology and Culture. He is co-author (with Helena Norberg-Hodge and John Page) of *From the Ground Up: Rethinking Industrial Agriculture. Address:* c/o ISEC, 850 Talbot, Albany, CA 94706, USA.

Edward Goldsmith is the publisher and founding editor of *The Ecologist* magazine, and author of numerous articles and books, including *The Way: An Ecological World-View*, and *The Great U-Turn*. His *Blueprint for Survival*, published in 1972, has been translated into 17 different languages. *Address:* c/o The Ecologist, Station Road, Sturminster Newton, Dorset DT10 1BB, UK.

Nicholas Hildyard is co-editor of *The Ecologist* magazine, and co-author (with Edward Goldsmith) of *The Social and Environmental Effects of Large Dams*, and *5000 Days to Save the Planet*. *Address:* c/o The Ecologist, Station Road, Sturminster Newton, Dorset DT10 1BB, UK.

Evelyne Hong is editor of *Third World Resurgence*. She is the author of *Natives of Sarawak*, and *See the Third World While It Lasts: A Case Study of Tourism in Malaysia*. *Address:* c/o Third World Network, 87 Cantonment Road, 10250 Penang, Malaysia.

S. M. Mohamed Idris is founder and president of both the Consumers' Association of Penang, and Sahabat Alam Malaysia (Friends of the Earth, Malaysia). He is coordinator of the Third World Network and is publisher and chief editor of the group's monthly magazine, *Third World Resurgence*. *Address:* c/o Third World Network, 87 Cantonment Road, 10250 Penang, Malaysia.

Martin Khor was formerly a professor of political economy, and is now Research Director of the Consumers' Association of Penang. He is also the Vice-President of the Third World Network, the Asia Pacific People's Network and Sahabat Alam Malaysia (Friends of the Earth, Malaysia). He is the author of numerous articles and several books, including *The Malaysian Economy: Structures and Dependence*. At present, he is particularly involved in issues relating to international trade. *Address:* c/o Third World Network, 87 Cantonment Road, 10250 Penang, Malaysia.

Sigmund Kvaløy is a farmer, writer and eco-philosopher. His work has been profoundly influenced by his close association with the Sherpa culture of Nepal. He has long been active in the movement to keep Norway out of the European Community. *Address:* Saetereng, 7496 Kotsoy, Norway.

Stephanie Mills is a writer on environmental issues, and has long been active in the Bioregional movement. She is the author of numerous papers and books, including *Whatever Happened to*

Ecology? and *In Service to the Wild*. She was formerly editor of *Not Man Apart*, the newsmagazine of Friends of the Earth.

NØFF is a coalition of Norwegian citizens which is working to prevent Norway from entering the European Economic Community. *Address:* c/o Natur och Ungdom, Torgg. 34, N-0183 Oslo, Norway.

Helena Norberg-Hodge is Director of the International Society for Ecology and Culture and its daughter organisation, The Ladakh Project. She has spent much of the last two decades in Ladakh ('Little Tibet') helping the Ladakhi people to maintain their cultural identity and ecological integrity in the face of rapid modernisation. She is the author of *Ancient Futures: Learning from Ladakh*, and co-author of *From the Ground Up: Rethinking Industrial Agriculture*. *Address:* c/o ISEC, 21 Victoria Square, Clifton, Bristol BS8 4ES, UK.

John Page, Programmes Director of the International Society for Ecology and Culture, has coordinated the technical and cultural programmes of the Ladakh Project for the last decade. He is the producer of the documentary video, *Ancient Futures: Learning from Ladakh*, as well as the video companion to *The Future of Progress*. *Address:* c/o ISEC, 21 Victoria Square, Clifton, Bristol BS8 4ES, UK.

Silvia Ribeiro works with Förlaget Nordan, a Swedish publishing cooperative. She is an active member of Friends of the Earth, Sweden, and has been involved in the management of Future Earth (see Birgitta Wrenfelt, below). *Address:* Elsa Beskows g.21, 1 tr. 126 66 Hagersten, Sweden.

Filipina del Rosario Santos is Acting Executive Director of the Appropriate Technology Centre for Rural Development in the Philippines, an NGO engaged in social programmes, research and education at the grassroots level. *Address:* c/o ATCRD, PO Box 7368, ADC-NAIA, Pasay City 1300, Philippines.

Vandana Shiva is coordinator of the Research Foundation for Science, Technology and Natural Resource Policy, in India. She is also actively involved in the Chipko ('tree-hugging') movement of the Himalayan foothills and co-chair of the Steering Committee of the Women's World Congress on the Environment. Her many publications include *Staying Alive: Women, Ecology and Survival in India*, and *Ecology and the Politics of Survival*. *Address:* c/o RFSTNRP, 105 Rajpur Road, Dehra Dun 248001, India.

Aman Singh is one of the leaders of Tarun Bharat Sangh, a Gandhian village development programme in Rajasthan, India. *Address:* 36/77 Kiran Path, Mansovar, Jaipur 302020, Rajastan, India.

Gary Snyder is a poet and writer concerned with issues relating to ecology and Buddhism. His numerous publications include *The Practice of the Wild*, and *Turtle Island*, for which he received the Pulitzer Prize.

Wiert Wiertsema works with Both Ends, a Dutch organisation which seeks to facilitate the work of citizens' groups involved in environment and development issues. *Address:* c/o Both Ends, Damrak 28-30, 1012 LJ Amsterdam, Netherlands.

Birgitta Wrenfelt teaches ecology at Biskops-Arnö in Sweden. She has worked for many years with Friends of the Earth, Sweden, and was one of the founders of Future Earth, an international network for social and ecological development. Future Earth supports projects aimed at promoting self-reliance in the South, and runs courses and seminars on ecological agriculture and appropriate technology. *Address:* Havsfruv. 29, 161 38 Bromma, Sweden.

Index

❖

The International Society
for Ecology and Culture

The International Society for Ecology and Culture (ISEC) is a non-profit organisation based in Bristol, England and Berkeley, California, USA. Its primary goal is to promote critical discussion of the foundations of modern industrial society, while at the same time examining the principles necessary for the emergence of more sustainable and equitable patterns of living.

ISEC is the umbrella organisation of the Ladakh Project, which for fifteen years has been undertaking a wide-ranging programme in the Himalayan region of Ladakh aimed at exploring alternatives to conventional development.

Also available from ISEC:

The Future of Progress, a 30-minute video companion to this book, based on interviews with Edward Goldsmith, Martin Khor, Vandana Shiva and Helena Norberg-Hodge. Excellent for classroom use. VHS £12/$20.

Ancient Futures: *Learning from Ladakh*, an award-winning documentary based on the book by Helena Norberg-Hodge. Describes the impact of economic development and moderisation on the ancient traditional culture of Ladakh, or 'Little Tibet'; relevant not only for other 'developing' parts of the world, but for the industrialised West as well. 60 minutes, VHS. £20/$30.

Please specify UK or US format. Add £2/$2 postage and handling.

ISEC
21 Victoria Square
Clifton, Bristol BS8 4ES, UK
Tel: 01179 731575
Fax: 01179 744853

ISEC
850 Talbot
Albany
CA 94706, USA
Tel & Fax: (510) 527 3873